# COMPUTERS IN
# NUMBER THEORY

## BOOKS OF INTEREST

WAYNE AMSBURY
*Structured BASIC and Beyond*

M. CARBERRY, H. KHALIL, J. LEATHRUM, and L. LEVY
*Foundations of Computer Science*

W. FINDLAY and D. WATT
*Pascal: An Introduction to Methodical Programming*

HAROLD LAWSON
*Understanding Computer Systems*

DAVID LEVY
*Chess and Computers*

D. LEVY and M. NEWBORN
*More Chess and Computers*

TOM LOGSDON
*Computers and Social Controversy*

I. POHL and A. SHAW
*The Nature of Computation: An Introduction to Computer Science*

DONALD D. SPENCER
*Computers in Number Theory*

# COMPUTERS IN
# NUMBER THEORY

## DONALD D. SPENCER

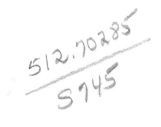

## COMPUTER SCIENCE PRESS

*Computer Science Press*
*11 Taft Court*
*Rockville, MD 20850 U.S.A.*

1 2 3 4 5 6                                                87 86 85 84 83 82

**Library of Congress Cataloging in Publication Data**

Spencer, Donald D.
    Computers in number theory.

    Includes index.
    1. Numbers, Theory of—Data processing. 2. Basic (Computer program language) I. Title.
QA241.S633     512′.7′02854     81-17452
ISBN 0-914894-27-7             AACR2

*To Laura*

# Preface

The computer is one of the most important calculating tools available to mathematicians. Today its use is quite commonplace, and for a mathematician, the ability to write and understand computer programs is becoming as necessary as the ability to use a hand calculator. Many complex mathematical problems, as well as those of a more routine nature, can be solved conveniently with a computer. Moreover, the search for solutions to research problems is greatly aided by computers. Many methods of calculation that are complex or tedious when done by hand can be accomplished with ease on a computer. Computational techniques such as iteration methods are feasible when a computer is used to accomplish the calculations. In fact, many of the useful methods of numerical analysis can be employed by a mathematician who knows computer programming and has the use of a computer system.

The purpose of this book is to introduce the reader to computer programming using number theory examples. The book can be used as a supplementary text in a college level number theory course or as a general interest book for anyone interested in computerized number theory.

The elementary theory of numbers has fascinated both professional mathematicians and laymen for more than 2000 years. Number theory concerns the most familiar of all numbers, the numbers we count with, and it may surprise you to realize how much you already know about them from experience. This book does not survey the whole area of number theory, but rather introduces many individual topics, some of which you may want to go into more thoroughly later. The chapters in the book are independent, although relationships are brought out.

The programming language used in this book is BASIC. Since BASIC is easy to learn and can be used with almost any computer system, it serves as a good introduction to computer communication. The student is encouraged to write computer programs to solve problems ranging from calculating prime numbers to generating magic squares. In this manner, the student can learn the programming and numerical techniques by application to number theory problems.

The first two chapters of this book introduce the reader to computers, problem solving with computers, the BASIC programming language, and number theory. The examples and discussions are directed toward number theory. This material is designed to provide readers with a working knowledge of the language so that they will be able to interpret BASIC programs and write their own programs. The remaining chapters in the book present basic numerical methods and the application of these methods to number theory problems that can be conveniently programmed. Actually, many of the methods presented can be used for calculation even when a computer is not available.

The book contains more than 80 BASIC programs to generate prime numbers, perfect numbers, amicable numbers, Armstrong numbers, Fibonacci numbers, magic squares, the greatest common divisor, and the least common multiple, and to obtain the solutions to many other number theory problems: the binary game of NIM, Chinese remainder theorem, Pythagorean triplets, Pascal's triangle, and others. The book contains a glossary of computer terms.

The BASIC programs in this book were executed on Apple II and Radio Shack TRS-80 microcomputers, and a Hewlett-Packard time-sharing system.

Donald D. Spencer
Ormond Beach, Florida
1982

# CONTENTS

# Chapter 1

# MEETING THE COMPUTER

**Preview**

This chapter is designed to give you a general overview of computer systems and to show you how they are used to solve problems. Number theory, algorithms, and flowcharts are also discussed in the chapter.

After completing this chapter, you should be able to:

1. Identify several application areas where computers are used.
2. Understand how computers can aid you in solving problems in number theory.
3. Describe the different types of computer systems.
4. List the basic steps in computer program development.
5. Write algorithms and draw flowcharts of computer solutions to problems.
6. Input simple programs into a computer system via a display/keyboard or printer/keyboard device.

## 1.1 IMPORTANCE OF COMPUTERS

The introduction of the computer in the last quarter century has changed the information needs of most people of our modern society. The computer is perhaps the most useful modern-day tool yet developed. Today we are using computers in ever-increasing numbers, in ways never imagined just a few years ago. They guide our astronauts to far away space stations and planets, they compute our bank accounts, they help engineers design bridges and airplanes, they count our votes, they control microwave ovens and automatic cameras, and they even help McDonald sell hamburgers and chocolate malts. Computers have radically altered the world of business. They have opened up new horizons to the fields of science and medicine, improved the efficiency of government, and changed the techniques of education. They have affected military strategy, increased human productivity, made many products less expensive, and lowered the barriers to knowledge. They are used to store information on credit, tax, insurance, financial, medical, buying habits, and so forth. Let us briefly explore some of the ways computers affect your life.

In the field of education, computers enable students and teachers to perform more effectively. With their assistance, persons may delve more deeply into areas that were once thought too time-consuming for hand calculation methods. Imaginative new devices and techniques have been developed to help students learn. As an example, students may sit at a typewriter-like device (called a terminal) and communicate directly with a computer.

In the field of medicine, computers are being used to help improve the health of people. They are used to help schedule hospital beds and to admit and discharge patients from hospitals. Some doctors use computers to help them diagnose an illness in a patient and to perform medical research.

Computers help engineers design automobiles, airplanes, bridges, and buildings. They help publishers produce newspapers, magazines, and books. Telephone companies use computers to route telephone calls.

The business world affects us all, whether or not we wish it to do so, and the business world functions with the assistance of computer systems. Business concerns use computers to maintain records, pay employees, keep track of inventory, control manufacturing, and perform hundreds of other functions.

A bank may record hundreds of thousands of transactions daily. Computers have become essential to the continued operation of banks. Rapid, accurate processing of large volumes of data is necessary in the banking business.

Retail businesses find more and more uses for computers. In some department stores a clerk using a cash register connected to a computer keys in a type-of-sale code, customer account number, merchandise price, and item inventory code. The computer quickly calculates the total sale and the status of the customer's account and displays the information.

In law enforcement, computers are used to help trace stolen goods, to study the causes of crime, to help locate missing persons and wanted fugitives, and to research methods of preventing crime.

Computers are used to reserve seats on airplanes, to help keep track of passenger baggage, and to help pilots fly airplanes safely. A computer aboard the giant Boeing 747 jetliner has been programmed actually to fly the airplane.

In the area of entertainment and novelty, computers have been programmed to play chess, compose original music, write poetry, and draw pictures.

These are only a few of thousands of examples of how computers help to keep a fast-paced society moving.

Computers can store every variety of information recorded by people and almost instantly recall it for use. They can calculate tens of millions of times faster than the brain and solve in seconds many problems that would take batteries of experts years to complete. No one should have to spend long hours adding endless columns of numbers, entering accounts in ledgers, keeping inventory records, or making out bills and checks. But this is all good and proper work for a computer. Beyond such mundane chores, of course, the computer does vital jobs that could never be done fast enough by unaided human minds.

Our lives are affected each day in some way by computers. In the future, the interaction between people and computers all over the world will increase. In time, computers will even respond to oral command and give reports in both written and spoken languages. Within the next two decades, it is estimated that most jobs will involve the use of computers, either directly or indirectly. It is important, then, that everyone understand something about computers and how they are used in our society.

Through space satellites and data communication links, all fields of information will some day be instantly available from computer centers around the globe and automatically translated into the language of the user. Although the computer is only slightly more than 30 years old, the field of computer science has grown so rapidly that it is now the largest industry in the world. This very moment, computers are running 24 hours a day in London, Tokyo, Rome, Paris, Moscow, Toronto, Mexico City, Amsterdam, Madrid, and Berlin, as well as in most American cities and towns.

## 1.2 A GLIMPSE OF NUMBER THEORY

Number theory is a branch of mathematics that deals with the natural numbers,

$$1, 2, 3, 4, 5, \ldots,$$

often called the positive integers.

Archaeology and history teach us that man began early to count. He learned to add numbers and much later to multiply and subtract them.

Dividing numbers was necessary to share a heap of pears or a catch of fish evenly. These operations on numbers are called calculations. The word calculation is derived from the Latin *calculus,* meaning a little stone; the Romans used pebbles to mark numbers on their computing boards.

As soon as men knew how to calculate a little, calculation became a playful pastime for many a speculative mind. Experiences with numbers have accumulated over the centuries, with compound interest, so to speak, and we now have an imposing structures in mathematics known as number theory. Some parts of it still consist of simple play with numbers, but other parts belong to the most difficult and intricate chapters of mathematics.

Number theory has many practical applications in engineering and physics, and has many uses in proving theorems in other fields of mathematics. An engineer who designs a gear train must use a form of number theory, and so must a physicist who undertakes an explanation of the interactions between atoms and radiation.

If you are a puzzle enthusiast, you can probably recall many puzzles that depend upon properties of numbers for their interest. If you have solved any of these puzzles, you have already been initiated into some of the elements of number theory. As a simple introduction to the field of number recreation, try baffling your nonmathematical friends with a mathematical trick. Tell your friends to follow these instructions:

> Think of a number
> Add 3 to this number
> Multiply your answer by 2
> Subtract 4 from your answer
> Divide by 2
> Subtract the number with which you started.

If your friend carries out these instructions carefully, the answer will always be 1, regardless of the number with which your friend started. We can explain why this trick works by using algebraic symbols, as shown below.

| | |
|---|---|
| Think of a number, | $x$ |
| Add 3: | $x + 3$ |
| Multiply by 2: | $2x + 6$ |
| Subtract 4: | $2x + 2$ |
| Divide by 2: | $x + 1$ |
| Subtract the original number, $x$: | $(x + 1) - x = 1$ |

If you like such games, you will find more in Chapter 9.

In recent years, computers have been used to aid mathematicians in the field of number theory. Two examples using the computer to solve number theory problems will be considered in this section. Many more examples are given throughout the book.

Perfect numbers are those whose divisors add up to the number itself. For example, 6 is a perfect number. Its divisors—1, 2, and 3—total 6. The next perfect number is 28—with divisors of 1, 2, 4, 7, and 14. Some 20 of these perfect numbers have been known for years, but anything beyond that was once thought to be unattainable. That was true until 1970 when a 18-year-old high school student, Roy Ferguson of Dallas, Texas, used a computer to produce the twenty-first perfect number, which contained 5688 digits. Although its precise determination required large amounts of computer processing time, the computer program used contained less than twenty-five individual instructions. Shortly after he found the twenty-first perfect number, Ferguson used another computer to compute the twenty-second and twenty-third perfect numbers. Computing these 5985- and 7723-digit numbers required about three hours on a large computer. The twenty-fourth perfect number, when expanded, has 12 003 digits.

On 23 May 1979, Harry Nelson and David Slowinski, computer specialists at the Lawrence Livermore National Laboratory in California, discovered a new prime number. A prime number is a number divisible only by itself and the number one. As numbers get larger, primes become harder to identify and increasingly rare. When written out, the new prime has 13 395 digits (Figure 1.1). Six students spent three hours just printing it on a blackboard. Without computer assistance it would be unthinkable to produce such large numbers. Naturally, I would like to print the whole number here, but due to lack of space in the book, you will have to settle for scientific notation. The number is $(2^{44497}) - 1$. Perhaps you would like to use your computer to find a larger one! This number took 300 hours to find on a CRAY-1 computer (see Figure 1.2) outputting one trillion bits per second, starting with the previous largest known prime number and trying all likely numbers. Clearly the new prime won't prove useful for measuring anything real. However, computer experts and cryptographers hope that it will provide keys to the complex codes needed to insure the secrecy of data banks.

Although these examples show the power of the computer, they also show that the honor of finding the largest perfect number, prime number, or any other special number will last only until someone else is willing to devote more computer time to the problem.

The emphasis throughout this book is on computer solutions to number theory problems. After we examine how computers may be used to help us solve problems (Chapters 1 and 2), we will use the computer to generate some

**Figure 1.1**    The newly discovered prime number goes on, and on, and on, and on—
for 13 395 digits.

special numbers as well as to solve many other interesting problems in
number theory (Chapters 3 through 9).

### 1.3  HOW TO RECOGNIZE A COMPUTER

A computer is an automatic device that performs calculations, makes decisions, and has the capacity for storing and instantly recalling vast amounts of
information. It processes information that can be represented in many
forms, including numbers, letters, words, sentences, sounds, pictures, for-
mulas, control signals, punctuation marks, and mathematical signs. The
processing is controlled by a set of step-by-step instructions called a computer program. A computer can do precisely those jobs for which we can
devise such a set of detailed instructions.

Regardless of differences in physical appearance, virtually every computer
may be envisioned as being divided into four logical units or sections.

1.  A central processing unit.
2.  A means of putting information into the system.
3.  A means of getting information out of the system.
4.  Facilities for storing information.

**Figure 1.2**    The CRAY-1 supercomputer was used to find a prime number with 13 395 digits. *Courtesy of the Department of the Air Force.*

The central processing unit, or CPU, is the heart of the computer. It is made up of two parts, the control unit, which coordinates the operations of the entire computing system, and the arithmetic/logic unit, which does the calculations. The CPU contains only a small amount of working storage (also called memory) to use when making calculations. The remaining storage, as well as the input and output devices by which the computer communicates with the outside world, are external to the CPU. In some computers a portion of main storage is actually part of the CPU.

Input and output units provide for communications with external equipment such as visual display devices, typewriters, and printers. The main storage of a computer is used to hold program instructions and the data that are being processed. Since main storage is expensive, lower cost auxiliary storage is often used to supplement the main storage of a computer.

### 1.3.1  Computer Storage

The computer's storage (or memory) is the part of the equipment that stores information for later use. Most computers have two types of storage: (1) main storage—a fast memory that is directly connected to the central processing unit (or, in some cases, an actual part of the central processing unit); (2) auxiliary storage—a slower storage that is used to supplement the main storage of a computer.

The main reason for the distinction between main and auxiliary storage is cost in relation to storage capacity and performance. Main storage must provide very fast performance and is much more costly than auxiliary storage devices per unit of capacity. Auxiliary storage must provide massive capacity for storing thousands, millions, or even billions of characters. Auxiliary storage devices are connected to the computer and may be accessed by the computer with little or no operator intervention. Auxiliary storage does not perform as rapidly as main storage, but is less expensive.

A new storage device, called magnetic bubble memory, is being used in many computers. The bubble memory provides medium speed storage at a price close to that of inexpensive auxiliary storage devices, but without either moving parts or the problems of reliability that moving parts entail. As bubble memory becomes more widely used, the distinction between main and auxiliary storage will tend to disappear. It will then be possible to store and rapidly access very large data files as if they were part of main storage.

The speed at which a computer can perform computations depends to a large extent on the performance of its main storage. Two types of main storage can be found in computers, semiconductor memories and magnetic core storage. Semiconductor memory is used widely in the newer computers, while magnetic core storage was popular in the older machines.

A semiconductor is an extremely small electronic component such as a transistor or diode. These components act like on-off switches. The direction of the electric current passing through each component, or cell, determines whether the position of the switch is on or off; that is, whether the bit is 1 or 0. In a semiconductor memory device, thousands of these miniature components are combined on a tiny silicon chip. Semiconductor memories are called integrated circuits, since all the necessary memory components are integrated on a single silicon chip.

A typical magnetic core storage unit is made up of thousands of tiny doughnut-shaped rings of a ferromagnetic material assembled on a cross-hatch of fine wires.

There are three commonly used auxiliary storage devices, magnetic disk, magnetic tape, and magnetic drum. Magnetic disk units and magnetic tape units are also used as input-output devices.

The most commonly used form of auxiliary storage device at present is magnetic disk. This type of device provides a large storage capacity at reasonable cost. A magnetic disk looks like a brown, grooveless phonograph record. Information is recorded on the disk in the form of magnetized spots.

Another type of disk unit that is used widely with small computer systems, especially microcomputer and minicomputer systems, is called a floppy disk (see Figure 1.3). In this sytem, information is recorded on a flexible disk called a diskette. To use the diskette either as a storage facility or as an input-output facility, one merely has to insert the diskette into a floppy disk drive.

Magnetic tape is a popular medium for transferring information. It is used not only as a method of getting information into and out of the computer, but also as an auxiliary storage. In magnetic tape equipment, information is represented in magnetized form on a long strip of magnetic tape.

Magnetic tape cassette units are used widely with microcomputer systems. The units use a magnetic tape that is loaded with a cassette. The cassette is

**Figure 1.3** Floppy disks are widely used with small computer systems, especially microcomputers and minicomputers. Shown here is a diskette being inserted in a floppy disk drive. *Courtesy of Verbatim Corporation.*

open at one end to permit insertion of magnetic read-write heads and the disk drive mechanism. Cassette units are basically simple, modest-performance devices that are relatively easy to operate and are inexpensive.

Magnetic drums are similar to magnetic disk units except that the recording surface is drum-shaped rather than disk-shaped. Their use is usually limited to special purposes in large-scale computer systems.

### 1.3.2  Input-Output Devices

For a computer to be useful, there must be convenient ways of putting the information to be processed into it, and getting the results out. That is the mission of input-output, or I/O equipment. Input devices feed information into the computer; output devices retrieve information from computer memory for human use.

Before information can be entered into the computer, it must be converted from a form that is intelligible to the user to a form that is intelligible to the computer. This conversion is accomplished by the input device. Input information is recorded on magnetic disks or tape as magnetized spots, on cards or paper tape as punched holes, on paper documents as line drawings or printed characters, and so on.

Output is data that has been processed by the computer. It may be in a form that can be directly understood by humans, or it may be retained in machine-readable form for future use by another machine. For example, an output device like the visual display can display information in a form understood by humans (see Figure 1.4). However, a magnetic disk unit used as an output device records information in a form that is useful only as input for further processing.

Some typical input-output devices are card reader, audio input, line printer, plotter, CRT, and plasma display.

Auxiliary storage devices such as floppy disk units, magnetic tape units, removable cartridge disk units, or magnetic tape cassette units may be used as both an auxiliary storage device and an input or output device.

### 1.3.3  Computer Systems

A computer system consists of a number of individual components, each of which has its own function. The system consists of equipment to send information to the computer (input devices), the central processing unit, storage devices, and equipment to accept information from the computer (output devices).

**Figure 1.4**  A visual display device. *Courtesy of Control Data Corporation.*

Complementing the hardware of a computer system is a software system that includes an operating system and other programs to aid us in preparing computer programs. The operating system is a program that controls the flow of jobs through the computer system. The user makes his or her request known to the operating system through system commands or control cards. The operating system then handles all details necessary to comply with the user's request.

Many years ago there was no problem in defining hardware and software. Hardware was the machine itself, the equipment. Software consisted of the programs and paperwork necessary to get the hardware to work. Today, however, the once clearcut distinction between hardware and software is blurring. Predesigned programs can be purchased nowadays not only in printed form, but recorded on all types of magnetic surfaces or even on plug-in modules or chips. When a plug-in module contains an unalterable program, it is considered dedicated. It actually becomes part of the hardware. Computer manufacturers are leaning in this direction to make the use of computers easier and therefore more appealing to many people and businesses. Hence, a new term firmware was defined. Firmware is software that is contained on what is normally considered hardware (chip or storage module). Several microcomputer systems use plug-in chips and modules.

Many people tend to think of computers as large, very expensive machines. Some computers do fit this description; however, many others do not. Computers can be classified into seven different groupings.

1. Supercomputers
2. Large-scale computers
3. Medium-scale computers
4. Small-scale computers
5. Minicomputers
6. Microcomputers
7. Microprocessors

## Supercomputers

A few businesses and organizations require extraordinary amounts of computing power. These include government agencies, scientific laboratories, aerospace companies, petroleum companies, research laboratories, and airline companies. The ever-increasing computational requirements of applications such as reservoir analysis in the petroleum industry, computer-aided design in the manufacturing aerospace industries, and seat reservation and ticketing in the airline companies will continue to require the processing power of supercomputers. Energy and power modeling are now a key part of the search for oil, for workable nuclear fusion, and for insuring nuclear reactor safety. Weather modeling is necessary for short-range forecasts and for long-range hazard predictions about atmospheric pollution. Such modeling requires computing at speeds approaching 100 million operations per second. As scientists modify their models in the 1980s, effective speeds well beyond 1000 million operations per second will be needed.

Supercomputers are the largest, fastest, and most expensive computers available. Supercomputers such as the Control Data CYBER 205 can generate results at a rate of 800 million results per second. A CRAY–1 supercomputer is shown in Figure 1.1. Supercomputers such as the CYBER 205 and CRAY–1 are being used by mathematicians and computer scientists to solve many complex number theory problems.

## Large-scale computers

A large-scale computer, together with its supporting equipment, will cost hundreds of thousands or millions of dollars. Computer systems of this size occupy very large rooms. Large systems can accommodate a large number and variety of supporting equipment. Large-scale computer systems are used primarily by government organizations and large corporations.

## Medium-scale computers

Medium-scale computer systems provide sufficient processing and storage facilities for many businesses and organizations. A typical system may cost several hundred thousand dollars or lease for up to $20 000* per month. A medium-scale NCR N–8450 system is shown in Figure 1.5.

## Small-scale computers

There is no doubt that the small-scale business computer will be a common sight in most small business firms—perhaps as commonplace as an office copier. The ever-increasing costs and complexities of doing business are forcing small businessmen to find new ways to cut their labor costs and gain tighter control over their operations, and a small computer system can help immeasurably in both these critical areas.

Small business systems are generally characterized by purchase prices in the $5 000 to $100 000 range.

## Minicomputers

Minicomputers have been around for about 15 years. These low-cost, compact, yet surprisingly powerful computers are being used in applications in business, education, and government. But what, exactly, is a minicomputer? The typical minicomputer costs about $20 000, uses integrated circuits, and is housed in a compact cabinet suitable for either tabletop use or for mounting in some type of cabinet. It weighs about 22.5 kilograms (50 pounds) and requires no special air conditioning. A minicomputer system that includes supporting equipment can cost as much as $100 000. A Digital Equipment Corporation PDP–11/44 minicomputer system is shown in Figure 1.6.

## Microcomputers

A microcomputer is a relatively inexpensive computer. The Apple II microcomputer costs around $1 000 and is popular in schools, homes, and businesses (see Figure 1.7). The TRS–80 Color microcomputer system costs about $400 and is designed for general use (see Figure 1.8). The Motorola

---

*Following the recommendations of the American National Metric Council, commas will not be used to group digits in this text, since the comma has been traditionally used as a decimal marker in other countries.

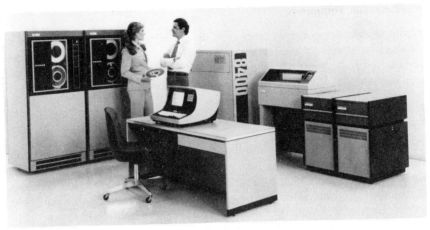

**Figure 1.5**    A medium scale NCR N-8450 computer system. *Courtesy of NCR Corporation.*

**Figure 1.6**    A PDP-11/44 minicomputer system. *Courtesy of Digital Equipment Corporation.*

MC6801 chip is a circuit that contains all functions of a computer. A microcomputer chip is often referred to as a computer on a chip.

Personal computers (broadly defined as microcomputer systems affordable by individuals and intended for personal rather than commercial use) are being used in thousands of application areas.

**Figure 1.7**   The Apple II microcomputer is widely used in schools, businesses, and homes.

Business microcomputer systems are just beginning to be accepted and used by small businesses. Microcomputer chips are widely used as the control units of personal and business microcomputers as well as other electrical devices.

Microcomputers have only been available for a few years. However, it is clear that they will have a greater effect on more people and businesses than any other kind of computer.

**Microprocessors**

A microprocessor is a semiconductor large scale integrated (LSI) circuit or very large scale integrated (VLSI) circuit that is essentially only a central processing unit; that is, it does not have significant amounts of on-chip memory and input-output logic.

Microprocessors are small enough and inexpensive enough so that they can be incorporated in other machines. Thus, many consumer products have built-in computers. These computerized products can accept and carry out far more complex operations than can their noncomputerized counterparts.

Microprocessors are used in video games, hand-held calculators, sewing machines, pinball machines, language translators, microwave ovens, cameras,

**Figure 1.8**   The Radio Shack TRS-80 Color Computer. *Courtesy of Radio Shack, a division of Tandy Corporation.*

automobiles, television sets, chess machines, washing machines, phototype-setting machines, gas station pumps, slot machines, paint mixing machines, and point-of-sale terminals.

### 1.3.4   Computer Trends

Computer technology has improved at a tremendously fast pace during the past three decades. During the past five years alone, computer equipment has become 100 times faster and 1000 times smaller and less expensive to operate. When computers were originally developed, many thought that only a few large businesses could use them; they were seen as too powerful, costly, and complicated for most concerns. Today, hundreds of thousands of microcomputers and microprocessors are used each day. The growth has been nothing short of phenomenal. At the same time, computers have gotten better and better.

It is interesting to compare the ENIAC (the first significant electronic digital computer) with today's microcomputer chip. The ENIAC occupied a space of 139.95 square meters (1500 square feet), weighed about 30 tons, contained about 19 000 vacuum tubes, and required 130 kW of power. ENIAC could perform 5000 additions per second. A microcomputer chip can perform calculations many times faster than the ENIAC could.

During the 1980s we can expect to see a continuation of the trend toward physically small, powerful, low cost computers.

## 1.4   TELLING THE COMPUTER WHAT TO DO

A computer is nothing more than a machine that follows detailed instructions given to it by you, the programmer. These instructions, collectively known as a computer program, are written in a specific computer language that the machine has been built to understand. BASIC is a carefully constructed English-like language used for writing computer programs. Instructions in the BASIC language are designed to be understood by people as well as by the computer. BASIC's popularity is evident when you consider that it is used extensively in thousands of educational institutions at all levels from elementary school to university graduate schools.

As indicated earlier, a computer program is a detailed set of instructions that directs the computer to solve a specific problem. The instructions in BASIC are both few in number and simple so that the task of learning BASIC is easy. Learning how to program is a different matter entirely. Although not difficult, it does require some thought as to the exact sequence in which instructions are to be executed.

Since a computer executes instructions sequentially, extreme care must be taken to put the instructions in the right order. The sequence of steps for the solution to any problem is known as an algorithm and must be determined by the programmer before writing the program. Don't let the term algorithm disturb you; it is simply another name for a plan of how to go about solving any type of problem. Actually, you use plans (or algorithms) for every task that you perform. For example, you don't need to plan or think out an algorithm for turning on a specific television program, yet the following steps are involved:

1.  Consult TV guide for station and time.
2.  Check watch for current time.
3.  If it is time for your show, turn TV set on.
4.  Set TV to proper channel.

5.  Adjust volume control.
6.  Adjust brightness and color.
7.  Sit down and watch the show.

As ridiculous as this example seems, it illustrates the fact that algorithms deal both with the individual activities that must take place and with the sequence or order of the activities. Since the computer executes instructions sequentially, it is absolutely essential that you set up these instructions in the proper order to get the job done.

How many of us have ever failed to be impressed by a magic square? The magic square shown here is arranged so that the sum of the numbers in any row, column, or diagonal is always 15.

|   |   |   |
|---|---|---|
| 8 | 1 | 6 |
| 3 | 5 | 7 |
| 4 | 9 | 2 |

This type of magic square can be generated as follows:

1.  Place the number 1 in the center cell of the top row.
2.  Assign three consecutive numbers in an oblique direction up and to the right. When this procedure carries the number out of the square, write that number in the cell at the opposite end of the column or row.
3.  After each group of three numbers has been assigned, move down one box and repeat rule 2.

This set of rules is called an algorithm. Simply stated, an algorithm is a recipe or list of instructions for doing something. If a computer is used to solve the problem, an algorithm provides the logical steps that the computer will follow in the solution.

A recipe in a cookbook is another example of an algorithm. The preparation of a certain dish is broken down into many simple instructions or steps that anyone experienced in cooking (and most who are not) can understand.

Almost everyone is familiar with the algorithm for placing a station-to-station telephone call. If not, you may find it in any city telephone directory.

1. Dial "1".
2. Dial the area code, if it is different from your local area code.
3. Dial the 7-digit telephone number for the place you want.
4. Give the operator your number, if you are asked for it.

Perhaps the best approach to use in developing an algorithm to solve a specific problem on the computer is to think in terms of "How would I do it without the computer?" Or, "How would I solve it by hand calculations, that is, with pencil and paper?" With few exceptions, the plan for doing the task by hand is likely to be the same as for doing it on the computer.

## 1.5 FLOWCHARTING

To help in the process of determining the correct sequence of instructions, most programmers use the technique of flowcharting. A flowchart is a graphical description of the logic, broken down into very simple steps. As an example, suppose we have a problem where we want to take the values of A, B, and C and add them together to reach the answer D. The flowchart for this problem would look like that shown below.

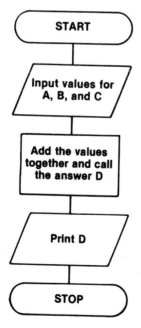

Although there are many different types and levels of flowcharts, several general rules should be followed in the preparation of all flowcharts, regardless of the complexity of the problem.

1. Use standard symbols.
2. Develop the flowchart so that it reads from top to bottom and left to right whenever possible. Use arrowheads to indicate direction.
3. Keep the flowcharts readable and simple. Leave lots of room between the symbols.
4. Write simple and descriptive messages in the flowchart symbols.
5. Direct only one line into a particular symbol.
6. Use only vertical and horizontal straight lines.

Most flowcharts can be prepared using four basic symbols. These symbols are linked sequentially through the use of connecting lines and directional arrows. The four symbols are as follows:

Terminal symbol—used at the beginning and end of a flowchart.

Input-output symbol—used when inputting data into the computer or when printing a result.

Decision symbol—used at a decision point in a problem.

Process symbol—used to specify a computation or operation.

Flowcharts may be drawn on paper of any size, although standard notebook paper size is usually preferred.

Flowcharts help you in many ways, but most importantly they force you to lay out the computer solution to a problem in a logical manner. The preparation of a clear, concise, and accurate flowchart is a crucial part of solving a problem on the computer (see Figure 1.9).

## 1.6  WRITING THE PROGRAM

Once a problem has been suitably formulated for a computer, by devising an algorithm and drawing a flowchart, it then becomes necessary to develop an explicit set of instructions expressed in some form suitable for comprehension by the computer. Such a plan for the solution of a problem by a computer, written in a suitable programming language, is called a program.

**Figure 1.9**  Flowcharts force one to lay out the computer solution to a problem in a logical manner. Shown here is a computer user drawing a flowchart. *Courtesy of IBM Corporation.*

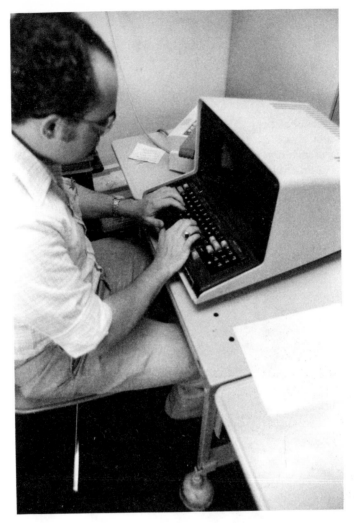

**Figure 1.10**  A mathematician using a computer terminal to help him solve a problem. *Courtesy of Northern Telecom Systems Corporation.*

Computer programs can be written in various programming languages. The simplest programming language is BASIC, which was developed at Dartmouth College in the 1960s. Other languages used for mathematical programming are FORTRAN, PL/1, APL, and Pascal.

**Review Exercises**

1. What is a computer?

2. Define "program."

3. Name the major classifications of computers.

4. What type of business might use a supercomputer?

5. Explain the difference between minicomputers, microcomputers, and microprocessors.

6. Explain how microcomputers might be used in future years.

7. List some consumer products that utilize microprocessors.

8. Modern computers are becoming small, reliable, fast, and cheap. Explain why these trends are important to computer users.

9. List the four basic parts of any computer system.

10. What is the purpose of input units? Of output units?

11. What is the main storage of a computer used for?

12. Name the two types of storage found in most computer systems.

13. What type of main storage is used on modern computers?

14. What are the primary devices used for auxiliary storage on a computer system?

15. Name several input-output devices and describe the purpose of each.

16. What is meant by hardware? Software? Firmware?

17. What is the purpose of a computer program?

18. What is an algorithm?

19. What is a flowchart?

20. What are some advantages to using a programming language such as BASIC?

21. What is a terminal?

# Chapter 2

# BASIC PROGRAMMING

**Preview**

This chapter is designed to show you how computer programs are written. First a few elementary concepts of the BASIC programming language are introduced. Then you are shown how these concepts can be used to write programs.

After completing this chapter, you should be able to:

1. Understand the most common statements in the BASIC language.

| | | | |
|---|---|---|---|
| REM | IF/THEN | END | RESTORE |
| LET | FOR | DIM | ON/GOTO |
| READ | NEXT | PRINT | |
| GOSUB | RETURN | STOP | |
| DATA | INPUT | GOTO | |

2. See how BASIC programming can be used to solve number theory problems.
3. Write BASIC programs to solve simple problems in mathematics and number theory.
4. Interpret simple programs written in BASIC.

## 2.1 BASIC PROGRAM ELEMENTS AND STRUCTURE

To many people, computers are the most powerful instruments of the twentieth century. A computer, though, is completely useless without a program, a set of instructions that tell it what to do. This applies to any computer, from a sophisticated supercomputer to a small microcomputer. In order to make it easy for people with different backgrounds and interests to use computers effectively, computer scientists developed symbolic programming languages.

24

For example, the FORTRAN language was designed for scientists and engineers, the COBOL language for business people, Pascal for computer scientists, and SNOBOL for language researchers. Another language, called BASIC (the letters B-A-S-I-C are an acronym for Beginner's All-Purpose Symbolic Instruction Code) was developed in 1964 by John Kemeny and Thomas Kurtz at Dartmouth College. This language was not designed for engineers, computer scientists, or businessmen, but for beginning computer users, whatever their interests might be. The BASIC language is easy to learn, easy to use, and easy to remember. BASIC consists of few syntax rules and a small number of statement types. It can be used in applications ranging from business to engineering.

BASIC is available on microcomputers, pocket computers, minicomputers, time-sharing systems, and small business systems, as well as on most larger computers. The language has been under continuous development, and several implementations of the language are available from manufacturers, schools, and service companies. New features have been added to the original BASIC language, and one frequently hears of "super BASIC," "simplified BASIC," and even "basic BASIC."

BASIC is one of the most widely used programming languages in business offices, homes, schools, industrial organizations, and professional offices. In education, BASIC is used in universities, colleges, and secondary schools. It is undoubtedly the best language for someone who may have occasion to use computers or who may want to learn how computers solve problems, but who does not want to become a professional programmer.

## BASIC Statements

Every statement in the BASIC language consists of some primary elements. The elements are shown as follows:

$$\text{ln} \quad \text{KEYWORD} \quad \text{PARAMETERS,}$$

where ln is a line number between 1 and 9999, a KEYWORD is a special word (such as LET, READ, PRINT, GOTO, etc.) that specifies the operation to be performed, and PARAMETERS further direct the operation to be executed. An example of a BASIC statement follows:

$$140 \quad \text{LET A} = 20$$

In this statement, 140 is the line number, the word LET is the keyword, and A, $=$, and 20 are parameters.

## Line Numbers

Every BASIC statement must appear on a separate line and must have a line number (or statement number). No two statements can have the same line number. Statements are executed by the computer in the order of their line numbers. Blank spaces may be inserted wherever desired in order to improve the readability of the statement. Thus, the statement

100    READA,B,C

is a valid statement and can be properly understood by the computer. However, a better way of writing this statement would be

100    READ A, B, C

Suppose we wished to use a computer to calculate the area of a circle using the formula

$$AREA = \pi R^2$$

given a value of the radius $R$. A BASIC program to solve this problem follows:

```
100    INPUT R
110    LET A = 3. 14159 * R↑2
120    PRINT R, A
130    END
```

We see that the program consists of four statements, each having its own line number. The line numbers increase successively from the first statement to the END statement. The statements contain the keywords INPUT, LET, PRINT, and END, respectively. The purpose of the first statement is to enter a numerical value for the radius (R) from a terminal connected to the computer. The second statement causes the values for the radius and area to be printed on the terminal. The last statement terminates the program. The symbol * is used to represent multiplication while the symbol ↑ means exponentiation.

When assigning line numbers you should leave several numbers between statements. A recommended method is to use 100 as the first statement, 110 for the second, 120 for the third, and so on. Thus space is available for inserting new statements if they are needed.

## Characters and Symbols

The allowable characters used in the BASIC language consist of the ten digits 0, 1, 2, 3, 4, 5, 6, 7, 8, 9; the 26 capital letters A through Z; and the following special characters:

| | | | |
|---|---|---|---|
| $ | currency | : | colon |
| + | plus | < | less than |
| − | minus | = | equal sign |
| * | asterisk | > | greater than |
| / | solidus (slash) | ≠ | not equal to |
| ↑ | up arrow | ! | exclamation mark |
| ( | left parenthesis | & | ampersand |
| ) | right parenthesis | , | comma |
| ' | single quote mark | ; | semicolon |
| " | double quote mark | # | pound |
| . | period or point | | blank (or space) |

A BASIC symbol is a series of one or more characters that has been assigned a specific meaning. Typical symbols are the minus sign (−) and the semicolon, which is used as a separator. A symbol consisting of more than one character is called a composite symbol, and it is assigned a meaning not inherent in the constituent characters themselves. Typical composite symbols are < = for "less than or equal to" and * for "multiplication." The symbols of the BASIC language are as follows:

| | | | |
|---|---|---|---|
| + | addition or prefix + | > | greater than |
| − | subtraction or prefix − | > = | greater than or equal |
| * | multiplication | < | less than |
| / | division | < = | less than or equal to |
| ↑ | exponentiation (or * *) | < > | not equal to (or ≠) |
| = | assigned to | = | equal to |
| . | decimal point | () | enclose group expressions |
| , | separator | " | used to enclose literals |
| ; | separator | | |

In some implementations of BASIC, lowercase letters can be used interchangeably with uppercase letters.

## Numeric Data

In BASIC, numbers can take the form of integers or fractional values in decimal form and can be positive or negative numbers. Some examples of valid BASIC numbers are:

|  |  |
|---|---|
| 61 | 0.0 |
| +82 | 7.006 |
| −234.7 | .54326107 |
| 0 | −.04 |
| 14623107 | −400 |

But numbers such as

$$2,610 \quad \sqrt{12} \quad {}^3/_4 \quad 6\,{}^1/_2 \quad \$40$$

cannot be used in BASIC. Fractions can, however, be represented in decimal form. For example, the mixed number 1¼ would be represented as 1.25. When numbers are very large or very small, it may be convenient to put them in scientific notation. When a number is placed in scientific notation it is reduced to a decimal number less than ten but greater than or equal to one times a power of ten. So, 100 in scientific notation would be $1.0 \times 10^2$. Since the keyboard of a terminal does not have superscripts, an E is placed between the decimal number and the power of ten. So,

1000 is the same as 1.0E3
10 000 is the same as 1.0E+4
−.0048 is the same as −4.8E−3

The number after the E (called the exponent) tells us how many places to move the decimal point. If the sign of the exponent is +, the decimal point is moved that many places to the right. If the sign of the exponent is −, the decimal point is moved that many places to the left. If in moving the decimal point we run off the end of the number, then we fill in with extra zeros as required.

## Character Strings

Sometimes it is desirable to work with nonnumeric information. Any collection of characters such as letters, letters and numbers, or letters, numbers, and special symbols is called a character string. For example,

### 427 HILLYBILLY AVENUE

is a string of 21 characters: 16 letters, 3 digits, and 2 blanks. In BASIC, strings are always enclosed in quotation marks. The maximum number of characters allowed in a string varies from computer to computer. The length of a character string is the number of characters between the enclosing quotation marks. The following examples show BASIC character strings:

> "NUMBER THEORY"
> "428-9160"
> "ODD-ORDER MAGIC SQUARE"
> "GCD FOR 10 NUMBERS"
> "GUESS THE LUCKY NUMBER"

Character string data are frequently used for printing descriptive information, such as column headings, messages, and identifiers.

### Names

Any data that is to be used in a program must be stored in the computer's storage, either prior to or during program execution. Computer storage consists of a large number of locations, each of which can hold a piece of data. These storage locations can be named by the program writer (programmer) and, subsequently, these names can be used to refer to the data stored in the locations. The program writer supplied names are called variable names. The variable names usually represent some numeric value, and because this value is represented by a variable name, it can be changed in value at the direction of the program. However, each variable name can represent only one value at a time.

In BASIC, a variable name may be either a single, alphabetic letter, such as

> B      R      U

or a single alphabetic letter followed by one numeric digit, like this:

> B6      S1      A7

If a variable is to contain a string, the second character must be a currency symbol (dollar sign) instead of a numeric digit. Thus, for example

> X$      A$      R$

are all valid alphanumeric names. Any of the storage locations thus named could be used to store information such as names, addresses, messages, etc.

Alphanumeric names can be assigned character strings through several BASIC statements (e.g., 20 LET X$ = "MONEY", and 400 READ A$, B$). String data may be assigned from one string variable to another (e.g., 20 LET A$ = X$). It may also be compared to another string variable (e.g., 20 IF (A$ = X$) THEN 400), and it may be output as a string variable (e.g., 100 PRINT G$).

When a variable name is chosen to represent a value, it is advisable to use a name that bears some resemblance to the value being stored. For example, to represent number, the variable N might be used, C might represent a counter, C$ might represent a price, and T could be used to represent a total. Care taken in naming variables in a program will pay large dividends in terms of readable and understandable programs.

### Beginning and Ending a Program

Suppose you wrote a program and put it away for use at some future date. When you pick up the program again, several months later, you may not remember what the program does, or why you wrote it in the first place.

To avoid this problem, it would be very convenient to be able to add some general information to the program to help jog your memory, or perhaps even tell you what the program is all about.

A BASIC statement, called a REMARK or simply REM statement, allows you to insert into your program whatever remarks you care to make. The format for this statement is

ln   REMARK comment<br>or<br>ln   REM comment

REM statements are not executed by the computer and may appear anywhere in the program. They offer you a convenient means to identify a program name, to call attention to important variable names, and to distinguish the major logical segments of a program. An example of a REM statement is

200   REM FIND 40 FIBONACCI NUMBERS

If a comment requires more than one line, a subsequent REM statement must be used. An example of this is

201   REM THIS PROGRAM<br>202   REM CALCULATES

    203    REM FIBONACCI AND
    204    REM PRIME
    205    REM NUMBERS

here we can see that the entire comment requires five lines.

Many program writers use REM statements to start each program. For example, the statements

    400    REM MAGIC SQUARE PROGRAM
    410    REM SAM WILSON
    420    REM JULY 10, 1982

identify the program (Magic Square Program), the program writer's name (Sam Wilson), and the date the program was written (July 10, 1982).

REM statements are often used to identify variables used in a program. For example, the following statements might be used to identify the variables in a program to compute compound interest.

    200    REM SUM OF 20 NUMBERS
    201    REM PROGRAM WRITER - JOHN JOHNSON
    202    REM K = COUNTER
    203    REM S = SUM
    204    REM A(K) = NUMBER ARRAY

Although the inclusion of a REM statement as the first line in a program is optional, the way a program concludes is clearly specified in BASIC. The END statement, although not executed by the computer, indicates that all statements in the program have been executed. It must be assigned the highest line number in the program. The END statement must be the last statement in a BASIC program. The general form of the END statement is

*ln*    END

Thus, the statement

    600    END

is a complete END statement. The use of an all 9s line number for the END statement is a fairly common programming practice. This convention serves as a reminder to the program writer to include the END statement and helps to ensure that it is positioned properly.

## System Commands

System commands are special commands to the computer giving instructions on what to do with a program. It should be remembered that these commands are not instructions used in the problem solution. System commands are always dependent on the particular system being used, with the command names sometimes differing. However, the command functions included here are common to many microcomputer and small business systems.

After logging in and giving the necessary password, or simply turning on the system if a microcomputer is being used, the system responds with a prompt: READY, >, [, ■, or some other similar symbol. This informs the computer user that the system is ready to accept BASIC program statements or system commands.

Assuming a BASIC program is to be written, the program statements when entered are placed in a temporary working area inside the computer's memory. Any program to be executed must be stored in this working area. Moreover, only one program at a time is allowed in this area.

To verify the fact that your BASIC program or program segment has been entered in the correct place, you need merely to type the system command LIST followed by a return. The system will now respond as shown below.

```
100    . . . . . . . .
110    . . . . . . . .
120    . . . . . . . .
130    . . . . . . . .
140    . . . . . . . .
150    . . . . . . . .
160    . . . . . . . .
170    . . . . . . . .
```

Statements can be deleted by typing the line number of the unwanted statement and following it with a return. For example, to eliminate line number 140, you would type

140 (and press return)

To prove that statement 140 has been deleted, just LIST your program.

```
100    . . . . . . . .
110    . . . . . . . .
120    . . . . . . . .
130    . . . . . . . .
```

150  . . . . . . . .
160  . . . . . . . .
170  . . . . . . . .

All system commands are activated as soon as you depress the return key. System commands do not have line numbers.

Another system command that you will be using frequently is RUN. This command says "Computer, please execute the statements in my program." The RUN command translates and executes the program in the computer's working area. If there are any syntax errors in a program they will be indicated at this time.

The NEW command clears the working area and makes the computer memory ready for a new program.

## 2.2 INPUT-OUTPUT

In BASIC, data may be entered as an integral part of the program or from a terminal during the execution of the program. The data input statements are the READ/DATA pair and the INPUT statement. The results of the execution of a program can be printed on the terminal by means of the PRINT statement.

### Getting Data into the Computer

In order for a computer to solve a problem, it must be provided with instructions telling it what to do, but also with data to use when carrying out the instructions. In BASIC, the READ/DATA statement pair and the INPUT statement may be used to supply data to a program. The DATA statement is used to create a data list, internal to the computer, and has the form:

*ln*   DATA data list

The data list consists of numeric values and character strings. Items in the data list must be separated by commas.

As an example, the statement

200   DATA 47

is a DATA statement containing one integer number, the value 47. This value would be read into the program and placed in some variable name by the READ statement. The statement

100    DATA 26, 48.3, .06, "BILL SMITH"

contains three numeric values and one character string. These values will be set up in a data list in a program, and each time a READ statement is executed, one value at a time will be taken from the data list for each variable specified in the READ statement. The DATA statement can be placed anywhere in the program before the END statement, however, it is good programming practice to place all DATA statements consecutively near the end of the program. More than one DATA statement can be used if desired. For example, the following statements

        100    DATA 64,23,17,19,86,99,22
        101    DATA 14,31,82,71,24,12,60
        102    DATA 55,75,15,43,29,18,62
        103    DATA 83,93,11,21,13,86,72

could be used to supply the necessary data to a program. Since one DATA statement was not adequate to contain all the data necessary, we continued to a second DATA statement, then to a third, and finally to a fourth statement.

The statement that causes data to be transferred from the DATA statement to a variable name within a program is the READ statement. The general form of the READ statement is

        *ln*    READ variable list

When the READ statement is executed, the values in the data list are assigned consecutively to the variables in the READ statement. Each READ statement causes as many values to be taken from the data list as there are variables in the READ variable list. The variable list consists of variable names separated by commas.

The following example illustrates the case of inputting both numeric and alphanumeric data.

        100    READ R$, B$, X, Y
        200    DATA "RED", "BLACK", 24, 12

Here the character string RED will be assigned to the variable R$, the character string BLACK will be assigned to the variable B$, the numeric value 24 will be assigned to the variable $X$, and the value 12 to $Y$.

In some cases the data values to be used in a program are not known beforehand and must be entered while the program is being executed. The INPUT statement allows the computer user to interact with an executing program and permits data values to be entered. When the computer encounters an INPUT statement, it accepts values for the variables that are part of the list in the INPUT statement provided by the user at a terminal. For example, in the statement

$$100 \quad \text{INPUT X, Y, Z}$$

the computer would halt execution of the program and wait for the terminal user to enter values for variables $X$, $Y$, $Z$. Once the user has entered these values, execution of the program continues.

The form of the INPUT statement is

$$ln \quad \text{INPUT variable list}$$

where the variable list contains variable names separated by commas. The INPUT statement causes the computer to print a ? and then wait for you to input data during execution.

### Getting Information Out of the Computer

To do even a simple problem, we need a way to get data into the computer and a method of printing the computed answer. Two ways of getting data into a computer were discussed in the last section. Results may be printed by using a PRINT statement.

The PRINT statement performs an important role in BASIC programming. It is through this statement that we are able to see the results of executing the program. The statement consists of a line number, the keyword PRINT, and a list of output items.

There are several forms of the PRINT statement. Appearing alone, the word PRINT causes a line to be skipped in the terminal output.

The PRINT statement can print literal data (messages), which can consist of letters, numbers and/or special symbols. For example, the statement

$$10 \quad \text{PRINT "28 IS A PERFECT NUMBER"}$$

will cause the computer to print on the terminal whatever is between the quotation marks. The above statement would cause the computer to print:

$$28 \text{ IS A PERFECT NUMBER}$$

A PRINT statement containing quotation marks is the only statement in BASIC in which blanks are counted. The computer will print the message enclosed between quotation marks exactly. For example, the statement

130   PRINT "TYPE     THE          NUMBER"

will cause four spaces to appear between the words TYPE and THE, and six spaces to appear between the words THE and NUMBER. In each case, the skipped spaces precisely match the skipped spaces in the statement.

The computer prints information exactly as it is written in the program. Here are two examples:

```
100 REM SAMPLE PRINT PROGRAM
110 PRINT "GOOD AFTERNOON HUMAN"
120 PRINT "PUNCH MY KEYS TO FIND"
130 PRINT "OUT WHAT I CAN DO"
140 END

RUN

GOOD AFTERNOON HUMAN
PUNCH MY KEYS TO FIND
OUT WHAT I CAN DO


100 REM PROGRAM WRITTEN BY
200 PRINT "ROGER WILSON"
300 PRINT "6 FT, 4 IN; 180 LBS"
400 PRINT "NET WORTH - $4.06"
500 END

RUN

ROGER WILSON
6 FT, 4 IN; 180 LBS
NET WORTH - $4.06
```

One form of the PRINT statement has the same characteristics as the READ and INPUT statements. Here is an example:

100   PRINT X, Y, Z

Notice the presence of a line number, the word PRINT, and the variables, X, Y, and Z separated by commas. This is a difference between the PRINT statement and the two input statements, however, the location in a print list must have had values stored in them before the PRINT statement is executed. If not, the computer outputs values of zero.

Variables and messages in PRINT statements must be separated by either a comma or a semicolon. Normally, the PRINT statement is used in a program as follows:

400   PRINT A,B,C

and a result such as

106      312      463

is produced (assuming that the value of A was 106, B was 312, and C was 463). When it is desired to keep the printout close together, a semicolon is used as a separator. For example, the statement

10   PRINT A; B; C

might cause the following output to be printed:

106    312    463

In most computer systems, the comma sets the spacing at a field width of fifteen spaces, and the semicolon places two blank spaces between the printed output. With seventy-five character spaces to a line, the use of the comma permits up to five fields of fifteen spaces for output.

Messages and variables can also be mixed in the same PRINT statement. For example, the statement

180   PRINT "AVERAGE COST =", C

will cause the message AVERAGE COST = to be printed in zone 1, and the value of $C$ to be printed in zone 2. Thus, if the current value of $C$ were 98, then the following would be printed:

AVERAGE COST = 98

## Running Programs

Now you have enough information about BASIC to write a program that could actually be run on a computer. To help you toward this goal, look at the following examples of complete BASIC programs and see if you can determine what the output for each would be:

### Example 1

```
200 REM SAMPLE PROGRAM
210 READ N$, A$, R$, W$
220 PRINT N$,A$
230 PRINT R$, W$
240 DATA "MARY AMUILLER", "42 BIGFOOT AVE"
250 DATA "TOM EVANS", "163 BEACH STREET"
260 END
```

### Example 2

```
110 REM SAMPLE PROGRAM
110 READ X$, S1
120 READ Y$, S2
130 PRINT "TEAM: "; X$, "SCORE: "; S1
140 PRINT "TEAM: "; Y$, "SCORE: "; S2
150 DATA "YANKEES", 4,  "REDS", 3
160 END
```

### Example 3

```
400 REM SAMPLE PROGRAM
410 PRINT "ENTER X, Y, Z"
420 INPUT X, Y, Z
430 PRINT X, Y
440 PRINT Z
450 END
```

Now, why not type one of these programs into a computer and have it execute the program. You will need to find out how to gain access to the time-sharing computer or microcomputer system at your institution. With most microcomputer systems you merely have to turn the machine ON and it is ready to accept a program written in BASIC. Microcomputers such as the Radio Shack TRS-80, Apple II, Commodore VIC-20, ATARI 800, and Texas Instruments 99/4A are designed to simplify the running of BASIC programs.

Most time-sharing systems are also very easy to use and one needs only to find out the proper procedure.

Let us now see how a computer is used to solve a simple BASIC program. The computer used in this example was a Radio Shack TRS-80 microcomputer. The operating procedure is similar on most other computer systems.

**STEP 1.** Get the computer ready to execute programs typed in the BASIC language. In the case of the TRS-80 we merely had to turn the machine ON.

**STEP 2.** Type the following program.

```
100 REM DIE ROLL
110 READ W, X, Y, Z
120 PRINT "ROLL THE DIE"
130 PRINT "ROLL 1 ="; W
140 PRINT "ROLL 2 ="; X
150 PRINT "ROLL 3 ="; Y
160 PRINT "ROLL 4 ="; Z
170 DATA 4, 1, 6, 2
180 END
```

If you made a mistake typing the program, press the RETURN key and retype the entire line. Ignore any error messages that are printed.

**STEP 3.** To obtain a listing of the program, just type LIST. The computer will type back all the BASIC statements in the program. If you see something you don't like in one of the statements, type it over. The last version you type of a statement is what is stored in the computer's memory. All the other versions are erased.

**STEP 4.** You are now ready to witness the computer executing your BASIC program. Simply type RUN, and away we go. In this example the program is designed to print five lines of information. The computer would print out the results as follows:

```
ROLL THE DIE
ROLL 1 = 4
ROLL 2 = 1
ROLL 3 = 6
ROLL 4 = 2
```

**STEP 5.**  After you are finished using the computer, turn it OFF. The exact steps will depend upon which computer you are using.

## 2.3  ASSIGNING VALUES TO VARIABLES

In the previous section, we studied how to assign values to variable names with the INPUT statement and the READ/DATA statement combination. Another way in which we can assign values to variables is with the LET statement, which is discussed in this section.

### Arithmetic and Alphanumerical Operations

In BASIC, the symbols used to indicate some mathematical operations are different than we have learned in mathematics classes. For example, the symbol for multiplication we learned in elementary arithmetic was an $\times$, but in BASIC the symbol for multiplication is a $*$. So five times eight would be written as $5*8$ in BASIC. The following chart shows the difference between the arithmetic operators in mathematics and in BASIC.

| Operation | Mathematics Symbol | BASIC Symbol |
|---|---|---|
| Addition | $+$ | $+$ |
| Subtraction | $-$ | $-$ |
| Multiplication | $\times$ | $*$ |
| Division | $\div$ | $/$ |
| Exponentiation | $x^2$ | $\uparrow$ or $(**)$ |

Consider $(x + y)^2$ as written in mathematical form. In BASIC $(x + y)^2$ is equivalent to $(X + Y) \uparrow 2$. The arrow $(\uparrow)$ is used since most computer terminals have neither raised nor subscript symbols.

Several kinds of BASIC statements may contain expressions, which are written something like algebraic expressions. They cause the current values of the specified elements to be combined in the specified ways. An element in an expression may be a variable, such as $X3$, or a constant, such as 206.

The computer normally performs its operations according to a hierarchy of operations. The system will search through an arithmetic expression from left to right and do specific operations according to the following pattern:

1.  Exponentiation
2.  Multiplication and division
3.  Addition and subtraction

One can alter this order of operations by using parentheses in the formula. The parentheses have no effect on the formula itself other than to direct the order of operations. The computer will always find the innermost set of parentheses and evaluate the part of the formula it finds according to the hierarchy of operations.

Two operation symbols must not be used in succession unless separated by parentheses. Thus the incorrect expression, $Z = X* - Y$, should be written $Z = X*(-Y)$.

A few examples of putting mathematical expressions into their BASIC equivalents are shown below.

| Mathematical Representation | BASIC Representation |
|---|---|
| $(a + b) \div c$ | (A + B)/C |
| $ab - xy$ | A * B − X * Y |
| $(a + b) \div c^2$ | (A + B)/C ↑ 2 |
| $((a \div b)c)^2$ | ((A / B) * C) ↑ 2 |
| $x^5$ | X ↑ 5 |

Arithmetic expressions appear in many forms, but the final result is the assignment of a constant, a variable, or some expression composed of constants and variables to some specified variable. This is accomplished in the BASIC language through the LET statement.

### The LET Statement

The LET statement is a statement that computes an arithmetic expression and assigns the result of that computation to a variable. The general format of the LET statement is:

$$ln \quad \text{LET variable} = \text{arithmetic expression}$$

A common use for the LET statement is to assign some initial value to a variable that is quite often used by other statements within the program. Here are a few examples:

```
200   LET Y = 4
300   LET W = 400
160   LET S = A
240   LET A = 1000
```

In the first example, statement 200 causes the value 4 to be assigned to the variable $Y$. This takes the place of any previous value which $Y$ may have had, and this value will never change unless $Y$ is used to the left of the "replaced by" sign ($=$) in another LET statement or in an INPUT or READ statement. Thus, $Y$ may be used for comparison purposes in an IF statement or as part of an arithmetic expression in another LET statement, and it will not alter its value. Similarly, in statement 300, the variable $W$ is assigned the value 400. In statement 160, the variable $S$ is assigned the current value of variable $A$, and in statement 240, the variable $A$ is assigned the value of 1000.

Shown below are several examples of valid BASIC assignment statements involving arithmetic expressions.

LET C = A * B
LET R = X + Y + Z
LET W = D * C − 36
LET X = Y − Z
LET A = C * D
LET V = A * R − 426
LET B = A + B * 43

The LET statement can also be used to assign string data to a variable. Thus, for example, the expression

100   LET X$ = "BLAISE PASCAL"

causes the character string BLAISE PASCAL to be assigned to the variable named X$. The expression

200   LET Y$ = "26 EAST GRANADA  AVENUE"

causes this address to be assigned to Y$.
    The expression

400   LET A$ = Y$

causes the contents of the variable Y$ to be assigned to the variable A$. After this operation, A$ and Y$ will contain exactly the same information.

## Illustration of Simple BASIC Programs

*Arithmetic Mean*

The arithmetic mean of a set of $n$ numbers, $X_1, X_2, X_3, \ldots, X_n$, is denoted by $\overline{X}$ (read "X-bar") and is defined as

$$\overline{X} = \frac{X_1 + X_2 + X_3 + X_4 + \ldots + X_n}{n} = \frac{\sum_{i=1}^{n} X_i}{n}$$

The grades of a student on six examinations were 84, 91, 72, 68, 87, and 78. The following BASIC program computes the arithmetic mean of the student's grades.

```
100 REM STUDENT GRADE PROGRAM
200 LET N = 6
300 READ X1, X2, X3, X4, X5, X6
400 LET B = (X1 + X2 + X3 + X4 + X5 + X6)/N
500 PRINT "AVERAGE GRADE ="; B
600 DATA 84, 91, 72, 68, 87, 78
700 END

RUN

AVERAGE GRADE = 80
```

*Average Daily Sales*

Bill Pickrell, a salesman for the Southern Canning Company, uses the following program to compute his total weekly sales and daily sales average. His input to the program is his actual daily sales figures.

```
100  REM  AVERAGE DAILY SALES
110  PRINT "SALES FOR MONDAY IS";
120  INPUT M
130  PRINT "SALES FOR TUESDAY IS";
140  INPUT T
150  PRINT "SALES FOR WEDNESDAY IS";
160  INPUT W
170  PRINT "SALES FOR THURSDAY IS";
```

```
180   INPUT A
190   PRINT "SALES FOR FRIDAY IS";
200   INPUT F
210   LET T = M + T + W + A + F
220   PRINT "--------------------------"
230   PRINT "TOTAL WEEKLY SALES ARE $"; T
240   LET S = T/5
250   PRINT "AVERAGE DAILY SALES ARE $"; S
260   END


RUN


SALES FOR MONDAY IS?600
SALES FOR TUESDAY IS?340
SALES FOR WEDNESDAY IS?510
SALES FOR THURSDAY IS?780
SALES FOR FRIDAY IS?490

--------------------------
TOTAL WEEKLY SALES ARE $2720
AVERAGE DAILY SALES ARE $544
```

*Compound Interest*

Assuming a given amount ($A$ dollars) is invested initially, at $R$ percent interest, for $N$ years, the total amount $P$ accumulated at the end of the period will be

$$P = A \cdot (1 + R)^N$$

The interest earned in this period is the difference between $P$ and $A$. The following program computes the compound interest for $A = 3000$, $R = .09$ and $N = 4$.

```
100 REM COMPOUND INTEREST
110 READ A, R, N
120 LET P = A * (1+R)↑N
130 LET I = P-A
140 PRINT "PRINCIPAL = "; A
150 PRINT "INTEREST RATE = "; R
160 PRINT "YEARS = "; N
170 PRINT "COMPOUND PRINCIPAL IS "; P
```

```
180 PRINT "COMPOUND INTEREST IS "; I
190 DATA 3000, .09,4
200 END

RUN

PRINCIPAL =  3000
INTEREST RATE =  .09
YEARS = 4
COMPOUND PRINCIPAL IS  4234. 74
COMPOUND INTEREST  IS  1234. 74
```

*Area of a Ring*

The area of a ring (or washer) is computed from the formula

$$A = \pi(R^2 - r^2)$$

where $A$ is the area, $\pi$ is 3.14159, $R$ is the outside radius, and $r$ is the inside radius.

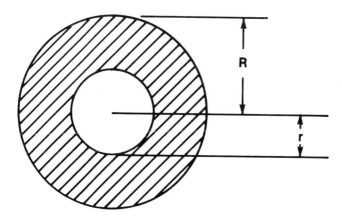

The following BASIC program uses the INPUT statement to input values for $R$ (R1) and $r$ (R2) at the time of program execution. (In this example $R = 8$ and $r = 3$.)

```
100 REM RING AREA PROGRAM
110 PRINT "OUTSIDE RADIUS IS";
120 INPUT R1
130 PRINT "INSIDE RADIUS IS";
```

```
140 INPUT R2
150 LET A = 3.14159*(R1↑2-R2↑2)
160 PRINT "AREA OF RING IS"; A
170 END

RUN

OUTSIDE RADIUS IS ? 8
INSIDE RADIUS IS ? 3
AREA OF RING IS 172. 787
```

## 2.4   CONTROL STATEMENTS

The sequence in which statements are executed during the operation of the program is ordinarily determined by the order in which the statements are physically arranged. Thus far we have discussed only programs that are composed of statements to be executed in an unchanging sequence. It is often desirable, however, to modify the sequence of program execution in relationship to some computed value or input value. There are several statements in BASIC that allow the program designer to specify the sequence of program execution. Thus, alternating program paths may depend on a specific condition at the time of execution. Line numbers serve as markers by which control statements can direct the sequence of a program. The program control statements discussed in this section are the GOTO statement, the IF/THEN statement, and the FOR/NEXT statement combination.

### The GOTO Statement

The simplest BASIC statement for altering the sequence of execution is the GOTO statement. This statement has the general form

*ln*    GOTO    line number

Suppose, for example, that you want to print the message

### THE PYTHAGOREAN TRIANGLE

many times. The following program prints this message and keeps returning control to statement 100 where the message is printed again. When will this program stop printing the message? Well, let's run the program and see.

```
100 PRINT "THE PYTHAGOREAN TRIANGLE"
200 GOTO 100
300 END

RUN

THE PYTHAGOREAN TRIANGLE
THE PYTHAGOREAN TRIANGLE
THE PYTHAGOREAN TRIANGLE
THE PYTHAGOREAN TRIANGLE
THE PYTHAGOREAN TRIANGLE
THE PYTHAGOREAN TRIANGLE
THE PYTHAGOREAN TRIANGLE
THE PYTHAGOREAN TRIANGLE
THE PYTHAGOREAN TRIANGLE
THE PYTHAGOREAN TRIANGLE
THE PYTHAGOREAN TRIANGLE
THE PYTHAGOREAN TRIANGLE
THE PYTHAGOREAN TRIANGLE
THE PYTHAGOREAN TRIANGLE
THE PYTHAGOREAN TRIANGLE
```

As you might guess, the program would continue printing the message until the terminal was worn out.

The following program has no practical importance other than illustrating the GOTO statement. The line numbers are executed in the following order: 210, 250, 230, and 270, resulting in the printing of the message COMPOSITE NUMBERS CAN BE FACTORED INTO TWO OR MORE SMALLER FACTORS.

```
200 REM COMPOSITE NUMBERS
210 PRINT "COMPOSITE NUMBERS"
220 GOTO 250
230 PRINT "INTO TWO OR MORE"
240 GOTO 270
250 PRINT "CAN BE FACTORED"
260 GOTO 230
270 PRINT "SMALLER FACTORS"
280 END

RUN
```

```
COMPOSITE NUMBERS
CAN BE FACTORED
INTO TWO OR MORE
SMALLER FACTORS
```

The program shown below illustrates how the LET statement can be used as either a counter of items or as an accumulator of values. The variable $K$ in statement 150 is being used to count the number of times the program cycles through the GOTO loop (statement 140 through statement 180). Each time statement 150 is executed, a constant value of 1 is added to the prior value of $K$. Statement 110 sets the starting value of $K$ to zero. Variable $B$ is used in the program to accumulate the sum of the values contained in the DATA statement. The purpose of statement 120 is to set the initial value of the variable $B$ to zero so that it can be used to accumulate the values of variable $A$ in statement 160 later on.

```
100 REM COUNTER EXAMPLE
110 LET K=0
120 LET B=0
130 PRINT "COUNTER", "A", "B"
140 READ A
150 LET K=K+1
160 LET B=B+A
170 PRINT K, A, B
180 GOTO 140
190 DATA 70, 50, 80; 95, 63
200 END

RUN
```

| COUNTER | A | B |
|---------|-----|-----|
| 1 | 70 | 70 |
| 2 | 50 | 120 |
| 3 | 80 | 200 |
| 4 | 95 | 295 |
| 5 | 63 | 358 |

On the sixth loop through the program the computer will realize that all the numbers have been used and it will print the message "OUT OF DATA IN LINE 140", or some other similar message, and stop.

## The IF/THEN Statement

The IF/THEN statement allows program control to be altered on a conditional basis, depending on the value of a conditional expression. The general form of the IF statement is

> *ln*    IF expression relation expression THEN line number

Both expressions are evaluated and compared by the relation in the statement. If the condition is true, program control is transferred to the line number given after THEN. If the condition is false, program control continues to the next statement following the IF statement.

In the IF/THEN statement, the following six relations symbols are used to compare values.

| Symbol | Relation |
|--------|----------|
| < | less than |
| < = | less than or equal to |
| > | greater than |
| > = | greater than or equal to |
| = | equal to |
| < > | not equal to (or $\neq$) |

The line number following the word THEN may be the line number of any executable statement in the program.

Quite often we wish to compare variables to constants; thus, we might have the expressions

```
100   IF A = 10 THEN 450
250   IF X > 100 THEN 800
300   IF R1 < = 20 THEN 200
```

A program can take alternate courses as it solves a problem. For example, assume you want to know whether 20% of 5000 is a better return on an investment than 18% of 6000. A program to determine the answer is:

```
100   REM   INVESTMENT DECISION
110   LET A = .20 * 5000
120   LET B = .18 * 6000
130   IF A > B THEN 190
```

```
140   IF A = B THEN 170
150   PRINT "18 PERCENT OF $6000 IS BETTER"
160   GOTO 200
170   PRINT "CALCULATIONS ARE EQUAL"
180   GOTO 200
190   PRINT "20 PERCENT OF $5000 IS BETTER"
200   END
```

```
RUN
```

```
18 PERCENT OF $6000 IS BETTER
```

Alphabetic fields may also be used in the IF statement; thus, the statement

$$400 \quad \text{IF X\$ = A\$ THEN 210}$$

causes program control to branch to statement 210 only if the variable X$ contains the same character string as the variable A$.

We can also test for alphanumeric constants; thus the statement

$$600 \quad \text{IF R\$ = ``BILL'' THEN 250}$$

will cause control to branch to statement 250 if the variable R$ contains the character string BILL.

Looping, one of the most important techniques in programming, makes it possible to perform the same calculation on more than one set of data. A loop consists of the repetition of a section of a program, substituting new data each time, so that each pass through the loop is different from the preceding one.

Let us consider the case where we wished to perform a loop a specific number of times. A convenient way to do this is by setting a counter to keep track of the number of repetitions. We test when the counter has reached its ending value by an IF statement. The following program will find the sum of the first ten odd numbers,

$$1 + 3 + 5 + \cdots + 17 + 19$$

In the program, the variable $N$ holds the numbers we are summing, $S$ contains the sum, and $C$ is a counter.

```
100 REM SUM OF THE FIRST TEN ODD
110 REM INTEGERS
120 LET N = 1
130 LET S = 0
140 LET C = 1
150 LET S = S+N
160 LET N = N+2
170 LET C = C+1
180 IF C <= 10 THEN 150
190 PRINT "SUM OF INTEGERS ="; S
200 END

RUN

SUM OF INTEGERS = 100
```

In the program we first set $S$ to zero in statement 130, and then, in statement 150, we increment its value by the current value of $N$. We make variable $N$ hold the successive odd numbers by starting with 1 and increasing it by 2 each time through the loop. Each time through the loop we increment the counter by one, and after the tenth time its value becomes eleven. The counter is then not less than or equal to 10, the test in statement 180 fails, and the branch to statement 150 is not taken. Hence, by executing statement 190, the program prints the answer, which is 100.

Considerable variety is possible in the way loops with counters can be written. We could have initialized $C$ at 10 in the previous program and decreased its value by one each time, testing for $C = 0$. We could have tested the counter before adding the value of $N$ to $S$, in which case we would terminate with a different test value. Such variations in the way in which programs can be written make the programming of computers interesting and challenging, but they also make it hard to understand a program written by someone else. Most experienced programmers use REM statements liberally to explain the purpose of each step of the program. Quite often they do this so they will themselves be able to recall what they were attempting to do in a program written some time ago.

The following program computes an average for a set of 10 numbers ($X$s).

```
100 REM  AVERAGE OF TEN NUMBERS
110 REM  K = NUMBER COUNTER
120 REM  S = SUM OF NUMBERS
```

```
130   REM   A = AVERAGE
140   LET K = 0
150   LET S = 0
160   REM   INPUT A NUMBER-VALUE OF X
170   PRINT "ENTER NUMBER"
180   INPUT X
190   LET S = S + X
200   LET K = K + 1
210   IF K < 10 THEN 180
220   REM   COMPUTE AVERAGE
230   LET A = S / K
240   PRINT "THE AVERAGE OF"
250   PRINT "TEN NUMBERS IS "; A
260   END


RUN

ENTER NUMBER
?6
?4
?3
?21
?10
?11
?6
?7
?8
?2
THE AVERAGE  OF
TEN NUMBERS IS 7.8
```

## The FOR/NEXT Statements

To do looping problems even more simply, BASIC provides two special statements to create a loop. These are the FOR and NEXT statements.

The FOR/NEXT statements are always used together and have the same general form.

In FOR $v = n_1$ TO $n_2$ STEP $n_3$
_____ BASIC statement_____
_____ BASIC statement_____
_____ BASIC statement_____
_____ BASIC statement_____
_____ BASIC statement_____
In NEXT $v$

where $v$ is a variable name acting as an index, $n_1$ is the initial value given to the index, $n_2$ is the value of the index when the looping is completed, and $n_3$ is the amount by which the index should be increased after each iteration through the loop (if the increment is 1, this portion of the statement may be omitted).

The FOR and NEXT statements can be used to simplify the formation of loops. Consider the following program.

```
100 FOR K = 1 TO 7
200 PRINT "EUCLID'S ALGORITHM"
300 NEXT K
400 END

RUN

EUCLID'S ALGORITHM
EUCLID'S ALGORITHM
EUCLID'S ALGORITHM
EUCLID'S ALGORITHM
EUCLID'S ALGORITHM
EUCLID'S ALGORITHM
EUCLID'S ALGORITHM
```

In this program, $K$ is the name of a counter. It has been set to vary from 1 to 6 in steps of 1. The program prints EUCLID'S ALGORITHM seven times. The loop consists of the three statements that begin with the FOR statement and end with the NEXT statement.

The next program prints the value of the counter as the loop is executed six times.

```
100 FOR X = 1 TO 6
110 PRINT "X ="; X
120 NEXT X
130 END

RUN

X = 1
X = 2
X = 3
X = 4
X = 5
X = 6
```

Suppose we wanted to write a program to determine the cube of each integer from 1 to 1000. Without using a loop, the program would be 1001 statements long and would look like this:

```
   1   PRINT "1"; 1 ↑ 3
   2   PRINT "2"; 2 ↑ 3
   3   PRINT "3"; 3 ↑ 3
   4   PRINT "4"; 4 ↑ 3

              ⋮

 999   PRINT "999"; 999 ↑ 3
1000   PRINT "1000"; 1000 ↑ 3
1001   END
```

To avoid the tedious task of writing the same statement a large number of times, we can write a simple program that will provide the same function; that is, build a loop that will count numbers 1 through 1000. The following program is an example of how to do this:

```
100 FOR X = 1 TO 1000
110 PRINT X, X↑3
120 NEXT X
130 END
```

In this program, statement 100 sets $X$ initially to 1, and a maximum value is established for use in the exit test. Statement 120 carries out two functions: $X$ is increased by 1 (modification), and the exit test is performed, that is, $X$ is compared to 1000. Notice that the value of $X$ is increased by 1 each time the computer goes through the loop. If you want a different rate of increase, you specify it by writing

$$100 \quad \text{FOR } X = 1 \text{ TO } 1000 \text{ STEP } 5$$

The computer would assign 1 to $X$ the first time through the loop, 6 to $X$ the second time through the loop, 11 on the third time through the loop, and 996 on the last time through the loop. Another step of 5 would take $X$ beyond 1000, so the program would proceed to the statement following NEXT after printing 996 and its cube. The STEP may be either positive or negative, and in the absence of a STEP instruction, the step size is assumed to be 1.

Loops can also be contained (nested) within other loops. The loops must actually be nested, and must not cross. A skeleton example is shown.

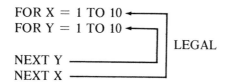

The following example of looping is not allowed.

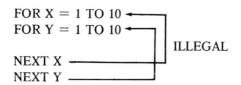

The program, ROOTS, which follows, illustrates the use of the nested loop. ROOTS is designed to print a table of square and cube roots. The two loops are nested. The outside loop is for the integers, $N$, of which we are to take the roots. The inside one is needed since for each $N$ we wish to compute two different roots. Here we use $R$ in the sense of the $R$th root, a square root and a cube root, all of which are written as the $1/R$ power of $N$.

```
100 REM PROGRAM ROOTS
110 PRINT "NUMBER", "SQ RT", "CU RT"
120 FOR N = 1 TO 12
130 PRINT N,
```

```
140 FOR R = 2 TO 3
150 PRINT N1(1/R),
160 NEXT R
170 PRINT
180 NEXT N
190 END
```

Consider the following problem. There are four scales to measure temperature: Fahrenheit (F), Celsius (C), Kelvin (K), and Rankine (R). Fahrenheit temperatures are converted to Celsius by subtracting 32° and multiplying the difference by 5/9. Kelvin temperatures are obtained by adding 273° to the Celsius reading. Rankine temperatures are obtained by adding 460° to the Fahrenheit reading. The following BASIC program computes temperatures equivalent to the following Fahrenheit temperatures: 45, 112, 89, 59, 73, 102, 36, 90, 27, 55, 65, 121, 34, 67, 97, 88, 25, 60, 17, and 44.

```
1000 REM TEMPERATURE CONVERSION
1010 PRINT "FAHRENHEIT", "CELSIUS", "KELVIN", "RANKIN"
1020 PRINT "----------------------------------------------------------- "
1030 FOR I = 1 TO 20
1040 READ F
1050 LET C = 5/9 * (F-32)
1060 LET K = C+273
1070 LET R = F+460
1080 PRINT F, C, K, R
1090 NEXT I
1100 DATA 45, 112, 89, 59, 73, 102, 36, 90, 27, 55
1110 DATA 65, 121, 34, 67, 97, 88, 25, 60, 17, 44
1120 END
```

RUN

| FAHRENHEIT | CELSIUS | KELVIN | RANKIN |
|---|---|---|---|
| 45 | 7.22222 | 280.222 | 505 |
| 112 | 44.4445 | 317.444 | 572 |
| 89 | 31.6667 | 304.667 | 549 |
| 59 | 15 | 288 | 519 |
| 73 | 22.7778 | 295.778 | 533 |
| 102 | 38.8889 | 311.889 | 562 |

| | | | |
|---|---|---|---|
| 36 | 2.22222 | 275.222 | 496 |
| 90 | 32.2222 | 305.222 | 550 |
| 27 | -2.77778 | 270.222 | 487 |
| 55 | 12.7778 | 285.778 | 515 |
| 65 | 18.3333 | 291.333 | 525 |
| 121 | 49.4445 | 322.444 | 581 |
| 34 | 1.11111 | 274.111 | 494 |
| 67 | 19.4444 | 292.444 | 527 |
| 97 | 36.1111 | 309.111 | 557 |
| 88 | 31.1111 | 304.111 | 548 |
| 25 | -3.88889 | 269.111 | 485 |
| 60 | 15.5556 | 288.556 | 520 |
| 17 | -8.33333 | 264.667 | 477 |
| 44 | 6.66667 | 279.667 | 504 |

After printing the heading, the program computes and prints four different values for each of the temperatures.

Have you ever wondered how high you could stack a pile of paper? Let's look at how a computer might do it. A sheet of paper is 0.5 mm thick. A stack of sheets of paper was started by laying down two sheets. The next addition to the stack was double the first, or four sheets. The third addition was double the second, or eight sheets. If this process were continued until 32 additions had been made, how high would the stack be? The following program determines that the stack of paper is 4.29497E + 09, or 4 294 970 000 millimeters high. This converts to 4 294 970 meters, or about 4294 kilometers high. This stack of paper is 1451 times as high as the Sears Tower, the tallest building in the world. The FOR/NEXT loop is found in the statements 160 through 200.

```
100 REM MOUNTAIN OF PAPER
110 PRINT "NUMBER", "HEIGHT"
120 PRINT "OF SHEETS", "IN MILLIMETERS"
130 PRINT
140 LET S = 2
150 LET A = 2
160 FOR D = 1 TO 32
170 PRINT S, S*.5
180 LET A = 2 * A
190 LET S = S + A
200 NEXT D
210 END
```

In the last section we used an IF/THEN in the computation of the average of ten numbers. Suppose we wished to compute the average of $N$ numbers, where $N$ could be any number. The following program requests that the user establish a value for $N$ and subsequently goes through a loop $N$ times entering the numbers to be averaged. Once the loop is completed the average is computed and printed. Type this program into your school's computer memory and use it to determine the average weight of the last 14 people you met, your average grade in some course to date, or the average points scored per game for the Dallas Cowboys 1981 football season.

```
100 REM AVERAGE A SET OF N NUMBERS
110 PRINT "TOTAL OF NUMBERS"
120 PRINT "TO BE AVERAGED IS";
130 INPUT N
140 LET S = 0
150 FOR K = 1 TO N
160 PRINT " NUMBER "; K;" IS";
170 INPUT R
180 LET S = S+R
190 NEXT K
200 REM COMPUTE AND PRINT AVERAGE
210 LET A = S/N
220 PRINT "THE AVERAGE OF THE "; N; " NUMBERS IS "; A
230 END

RUN

TOTAL OF NUMBERS
TO BE AVERAGED IS?8
NUMBER   1 IS?354
NUMBER   2 IS?897
NUMBER   3 IS?454
NUMBER   4 IS?970
NUMBER   5 IS?498
NUMBER   6 IS?372
NUMBER   7 IS?873
NUMBER   8 IS?231
THE AVERAGE OF THE 8 NUMBERS IS 581.13
```

Have you ever wondered how many different ways it is possible to make change for a dollar? The following program accomplishes this task.

```
100  REM  CHANGE FOR A DOLLAR
110  REM  PENNY(P), NICKEL(N), DIME(D)
120  REM  QUARTER(Q), HALFDOLLAR(H), COUNTER(C)
130  LET C = 0
140  FOR H = 0 TO 2
150  FOR Q = 0 TO 4
160  FOR D = 0 TO 10
170  FOR N = 0 TO 20
180  FOR P = 0 TO 100
190  IF (H * 50) + (Q * 25) + (D *
     10) + (N * 5) + P = 100 THEN 210
200  GOTO 220
210  LET C = C + 1
220  NEXT P
230  NEXT N
240  NEXT D
250  NEXT Q
260  NEXT H
270  PRINT "WAYS TO MAKE CHANGE F
     OR $1 00 = "; C
280  END
```

This program illustrates the nesting of five FOR/NEXT loops. The first coin values to add up to a dollar (100) is 100 pennies. When this occurs, the $C$ counter is increased by 1. $C$ is increased once more when there are 95 pennies and 1 nickel, then when there are 90 pennies and 2 nickels, then when there are 85 pennies and 3 nickels, and so on. The program determines that there are 292 different ways to make change for a dollar.

## 2.5  BASIC FUNCTIONS AND SUBROUTINES

The BASIC language makes available several built-in routines, called predefined functions, which accomplish desired objectives. For example, the square root of a number or the cosine of an angle can be calculated. Some tasks done on the computer require that a complicated operation or set of operations be repeated a number of times at different points in the program. BASIC makes user-defined functions and subroutines available. They provide an efficient way of handling such problems. Predefined and user-defined functions and subroutines are discussed in this section.

### Predefined Functions

The BASIC predefined functions can be grouped into four categories: trigonometric, exponential, arithmetic, and utility. Each function consists of a three-letter name followed by an argument enclosed in parentheses. The argument is a mathematical expression, which means it can include other functions.

The following list gives some mathematical expressions that include functions and their equivalent representation in BASIC:

| Mathematical Expression | BASIC Expression |
|:---:|:---:|
| $|X|$ | ABS(X) |
| $\sqrt{a^2 - b^2}$ | SQR(A↑2 − B↑2) |
| $\cos 30°$ | COS(30 * (3.14159/180)) |
| $\sqrt{1 - \sin^2 x}$ | SQR(1 − SIN(X) ↑ 2) |

### Trigonometric Functions

The BASIC trigonometric functions require the argument $X$ to be an angle measured in radians, you can multiply the number of degrees by .017453 or divide the number of degrees by 57.295780.

SIN(X)   The sine of $X$ is calculated.
COS(X)   The cosine of $X$ is calculated.
TAN(X)   The tangent of $X$ is calculated.
ATN(X)   The arctangent of $X$ is calculated (angle whose tangent is $X$).

In the following example,

$$100 \quad \text{LET A} = \text{SIN}(30 * .017453)$$

the sine of 30° is calculated and assigned to the variable $A$.

## Exponential Functions

For the following three exponential functions, the argument $X$ can be any expression.

EXP(X)  The natural exponent $e^x$ is calculated ($e = 2.718281\ldots$)
LOG(X)  The natural logarithm $\log_e X$ is calculated.
        $X$ must be greater than zero.
SQR(X)  The square root of $X$ is calculated. $X$ must be a positive value.

Consider the function SQR which takes the square root of a number. The square root of a number is the number which, when multiplied by itself, gives the original number. For example, the square root of 9 is 3 and the square root of 25 is 5.

$$SQR(36) = 6$$
$$SQR(60 + 40) = 10$$
$$SQR(70 * 5 + 50) = 20$$

The EXP function raises "e" to a given power. In the following example $e^{4.5}$ is seen to be 90.01713.

```
100  LET X = 4.5
200  LET A = EXP(X)
300  PRINT X,A
400  END
```

## Arithmetic Functions

For the following three arithmetic functions, the argument $X$ can be any expression.

ABS(X)  Determines the absolute value of $X$; $|X|$.
        ABS(64) = 64
        ABS($-64$) = 64
        ABS($-361$) = 361

INT(X)  Calculates the largest integer not greater than $X$.
        INT(12.6) = 12
        INT(41) = 41
        INT($-6.8$) = $-7$
        INT($-8$) = $-8$
        INT(.0006) = 0

SGN(X)   Determines the sign of $X$. The result is $+1$, $0$, or $-1$.

$$SGN(X) = 1 \quad \text{(where } X = 63.41)$$
$$SGN(X) = 0 \quad \text{(where } X = 0)$$
$$SGN(X) = -1 \quad \text{(where } X = -83)$$

The INT function can be used to round a number to the nearest integer. For example, the statement

$$200 \quad \text{LET A} = \text{INT(X} + .5)$$

can be used. If $X = 8.2$, then $X + .5 = 8.7$ and $INT(X + .5) = 8$, which is assigned to $A$. If $X = 4.8$, then $X + .5 = 5.3$ and $INT(X + .5) = 5$, which is assigned to $A$.

To round a number to the nearest tenth, the statement

$$100 \quad \text{LET B} = \text{INT(10} * \text{X} + .5)/10$$

can be used. If $X = 8.36$, then $10 * X + .5 = 84.1$ and $INT(10 * X + .5) = 84$; therefore $INT(10 * X + .5)/10 = 8.4$, which is assigned to $B$.

Business applications generally involve printing out the results of calculations in dollars and cents. Therefore, it is desirable to round off figures to the nearest cent. The following statement

$$300 \quad \text{LET C} = \text{INT(X} * 100 + .5)/100$$

can be used. Assume $X = 47.51798$, then

$$X * 100 = 4751.798$$
$$X * 100 + .5 = 4752.298$$
$$INT(X * 100 + .5) = 4752$$
$$INT(X * 100 + .5)/100 = 47.52$$

Therefore, 47.51798 was correctly rounded up to 47.52. As another example, suppose that $X = 84.14326$. Then

$$X * 100 = 8414.326$$
$$X * 100 + .5 = 8414.826$$
$$INT(X * 100 + .5) = 8414$$
$$INT(X * 100 + .5)/100 = 84.14$$

with the result that 84.14326 was correctly rounded down to 84.14.

Perform this computation yourself, supplying different values for $X$, until you understand why the formula works.

## Utility Functions

The following two functions are BASIC utility functions.

TAB(X)    Used in PRINT statements to tabulate output. $X$ specifies a print position. This function is discussed in Section 2.7.
RND(X)    Generates random numbers.

Many applications require the use of random numbers. The winner of a state lottery may be selected at random, a random sample of manufactured products may be inspected for possible defects, or tax returns may be inspected at random.

The RND function is useful to simulate random events, for example, flipping a coin or rolling a die. Of course, the computer cannot flip a coin or roll a die; however, it can be programmed to simulate a coin toss or a die roll. The following examples illustrate some uses of the RND function.

RND(0)    Will produce a number between 0 and 1, e.g. .543216
RND(2)    Will produce either a 1 or a 2. Can be used to simulate the flip of a coin, head = 1, tail = 2.
RND(6)    Will produce either a 1, 2, 3, 4, 5, or 6. Can be used to simulate the roll of a die.
RND(52)   Will produce a number in the range 1 to 52. Can be used to represent a card from a 52-card deck.

When the statement RANDOMIZE (some computers use RANDOM) is executed in a program before the RND function is used, it will initialize the random number generator to a new starting value. This will ensure that the RND function will produce a fresh sequence of random numbers that differs from any previous sequence. There are cases where you may wish to generate the same sequence of random numbers each time a program is run. This is a useful feature in situations where you are using random numbers to test your program.

The following four programs illustrate the RND statement.

**Example 1**    Random Numbers

The following program produces 28 random numbers between 1 and 900.

```
100 REM RANDOM NUMBERS
110 FOR X = 1 TO 7
120 LET A = RND(900)
130 LET B = RND(900)
140 LET C = RND(900)
150 LET D = RND(900)
160 PRINT A, B, C, D
170 NEXT X
180 END
```

RUN

| | | | |
|---|---|---|---|
| 684 | 109 | 229 | 784 |
| 849 | 548 | 216 | 217 |
| 561 | 252 | 50 | 436 |
| 735 | 801 | 112 | 628 |
| 563 | 210 | 539 | 857 |
| 424 | 179 | 881 | 189 |
| 73 | 501 | 416 | 2 |

## Example 2   Coin Tossing

If a coin is perfectly balanced, then the probability of tossing a head is equal
to the probability of tossing a tail. Hence, to simulate a coin-tossing game,
you simply generate random numbers and arbitrarily assign the occurrence
of the random number 1 to heads and 2 to tails. The following program will
cause the computer to simulate the flipping of a coin a specified number of
times.

```
100 REM COIN TOSSING
110 PRINT "TYPE THE NUMBER OF"
120 PRINT "COINS TO BE TOSSED"
130 INPUT C
140 PRINT "------------------"
150 FOR N = 1 TO C
160 LET R = RND(2)
170 IF R = 1 THEN 200
180 PRINT "TOSS"; N;" IS A TAIL"
190 GOTO 210
200 PRINT "TOSS"; N;" IS A HEAD"
```

```
210 NEXT N
220 END

RUN

TYPE THE NUMBER OF
COINS TO BE TOSSED?25
-------------------
TOSS 1   IS A TAIL
TOSS 2   IS A TAIL
TOSS 3   IS A TAIL
TOSS 4   IS A HEAD
TOSS 5   IS A HEAD
TOSS 6   IS A TAIL
TOSS 7   IS A HEAD
TOSS 8   IS A HEAD
TOSS 9   IS A TAIL
TOSS 10   IS A HEAD
TOSS 11   IS A HEAD
TOSS 12   IS A HEAD
TOSS 13   IS A HEAD
TOSS 14   IS A TAIL
TOSS 15   IS A HEAD
TOSS 16   IS A TAIL
TOSS 17   IS A TAIL
TOSS 18   IS A TAIL
TOSS 19   IS A HEAD
TOSS 20   IS A TAIL
TOSS 21   IS A HEAD
TOSS 22   IS A HEAD
TOSS 23   IS A HEAD
TOSS 24   IS A HEAD
TOSS 25   IS A HEAD
```

**Example 3**   Random Display

In this example we again generate a 1 or 2 randomly, but instead of assigning them to heads and tails, respectively, we print an asterisk ( * ) if a 2 occurs and a blank if a 1 occurs. This pattern of asterisks and blanks are contained within a rectangle specified by input parameters width (*W*) and height (*H*).

```
100 REM RANDOM DISPLAY
110 PRINT "ENTER WIDTH AND HEIGHT OF DISPLAY"
120 INPUT W, H
130 FOR I = 1 TO H
140 FOR K = 1 TO W
150 IF RND(2) = 2 THEN 180
160 PRINT " ";
170 GOTO 190
180 PRINT "*";
190 NEXT K
200 PRINT
210 NEXT I
220 END

RUN

ENTER WIDTH AND HEIGHT OF DISPLAY? 35, 20
    * * **  * ***** * ** * * ***
* ** * ****  *  ** * **** *   ***
  * *   ***   * *   ****  ****** *
 ** *** * ** * ***  ** **   **   *
  * *   *  ***** *   ****  *****
 ** * *  * *   * * * *******  *** *
 * *  * * * *  * ** * ** * ****
    ***     *  * ****** *  *   * *
 **  * *  *  * ****  *        *
 **   *    * *  * **   ** *  ** **
    * *  * **   **    **      * *
    *  * ** **   *****     *  *
 * *   **    *** * * *** ** * * *
    ***   * ** *** **  *      **
 *     *** * ** **  * * * * *******
   * *     ** * *** * * *  * *   *
  ***     * *   * * *  ***   * * *
 * ***** ** *** ***  *  ** *     *
 * ** * *   ******  *   ***    **
 *** ** ***         *** **    **
```

## User-Defined Functions

User-defined functions allow you to define up to 26 functions that are referenced in the same way as predefined functions. The definition of a user-defined function begins with keyword DEF, followed by the name of the function (either FNA, FNB, FNC, ..., FNZ), followed by the list of parameters enclosed within parentheses, followed by the function description.

If the description of computation of the function can be made in a single arithmetic expression, the list of parameters is followed by an equal sign followed by the arithmetic expression. For example, the following statement defines a function that converts inches to millimeters.

$$\text{DEF FNA(I)} = \text{I}/.04$$

If we include this definition in a program, we can use FNA to perform the inches-to-millimeter conversion wherever required in the program. We can code, for example:

$$100 \quad \text{PRINT FNA(23.7)}$$

instead of

$$100 \quad \text{LET Z} = 23.7/.04$$
$$200 \quad \text{PRINT Z}$$
$$\text{or}$$
$$100 \quad \text{PRINT } 23.7/.04$$

The function used in the following program triples any value given to it.

```
100 REM FUNCTION TO TRIPLE X
110 DEF FNS(X) = X + X + X
120 READ X
130 IF X = 999 THEN 180
140 LET A = FNS(X)
150 PRINT A
160 GOTO 120
170 DATA 3,14,20,33,999
180 END

RUN

    9
    42
    60
    99
```

As seen in the previous program (statement 110), the expression on the left side of the "replaced by" sign in a function definition specifies the function name (FNS) and the variable $X$. This variable is called a dummy variable, because it does not actually refer to a location in computer memory. It is used merely to hold the place of values to be inserted later. However, in this example, the dummy variable name ($X$) is also used to name the actual variable evaluated by the function at statement 140. The real heart of the function is the calculation to the right of the "replaced by" sign. This calculation is performed each time the function is called. It can be any arithmetic expression, including expressions that contain other functions. For example, the following user-defined functions are perfectly valid.

```
200   DEF FND(A,B,C) = (−B + SQR(B ↑ 2 − 4 * A * C))/(2 * A)
100   DEF FNB(X,Y) = X ↑ 2/Y ↑ 2 + ABS(X)
160   DEF FNW(X) = INT(X * 100 + .5)/100
```

## Subroutines

In large programs, it often becomes necessary to execute a particular set of instructions several different times, at different points in the program. Rather than having to repeat this set of instructions each time it is to be executed, BASIC provides two statements that allow the user to write a repeatable instruction set. The GOSUB statement is used to transfer program control to the first instruction in the set that is to be used. The RETURN statement is used as the last instruction in the set that is to be used. To illustrate, let us consider the following program segment:

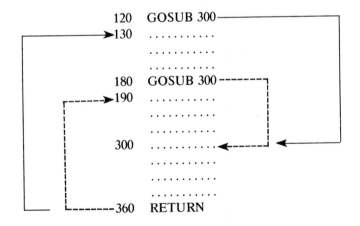

Statement 120 contains a GOSUB statement that will cause control to be transferred to statement 300, the first instruction in the set to be used. After executing the instructions following statement 300, statement 360 is encountered. This RETURN statement directs the computer to return control back to statement 130. A similar transfer and return operation would occur when statement 180 is executed. Here the RETURN statement at 360 would return control to statement 190.

The following program, which illustrates the use of the GOSUB and RETURN statements, computes the real roots of the equations

$$2x^2 + 12x + 3 = 0$$

$$2x^2 + 14x + 4 = 0$$

$$3x^2 + 20x + 2 = 0$$

$$2x^2 + 16x + 2 = 0$$

$$x^2 + 12x + 2 = 0$$

$$2x^2 + 10x + 1 = 0$$

We know from the quadratic formula that the two real roots of the equation

$$ax^2 + bx + c = 0$$

are

$$x = \frac{-b \pm \sqrt{b^2 - 4ac}}{2a}$$

Input to the program is six sets of values for a, b, and c.

```
100 REM ROOTS OF A QUADRATIC EQUATION
110 READ A, B, C
120 GOSUB 150
130 PRINT "ROOT 1 = "; X1," ROOT 2 = "; X2
140 GOTO 110
150 LET D = SQR(B↑2-4*A*C)
160 LET X1 = (-B+D)/(2*A)
170 LET X2 = (-B-D)/(2*A)
180 RETURN
```

```
190 DATA 2, 12, 3, 2, 14, 4, 3, 20, 2, 2, 16, 2, 1, 12, 2, 2, 10, 1
200 END

RUN

ROOT 1 = -.261268          ROOT 2 = -5.73873
ROOT 1 = -.298304          ROOT 2 = -6.7017
ROOT 1 = -.101376          ROOT 2 = -6.56529
ROOT 1 = -.126816          ROOT 2 = -7.87318
ROOT 1 = -.168736          ROOT 2 = -11.8313
ROOT 1 = -.101954          ROOT 2 = -4.89805
```

The subroutine itself is statements 150 through 180, which stores the computed roots in variables $X1$ and $X2$. The RETURN statement informs the computer that this is the end of the subroutine and to return to the next statement number following the GOSUB statement that caused activation of the subroutine. Every subroutine must have at least one RETURN statement in it. The GOSUB statement calls the subroutine each of the six times it is executed.

### 2.6 ARRAYS AND SUBSCRIPTS

There are many applications in which an arrangement of numbers is used over and over again. In BASIC, an arrangement or array of numbers is called either a list or a table.

When writing a program it is often convenient to refer to an entire collection of items at one time. Such a collection is called an array. For example we may be concerned with a list of items (also called a one-dimensional array), or with a table of values (known as a two-dimensional array). The BASIC language allows us to refer to the elements of lists and tables as though they were ordinary variables, thus making array manipulation as simple as possible.

The individual elements of a list or table are known as subscripted variables. Any such element can be referred to by stating the name of the list or table followed by the value of the subscript enclosed in parentheses. Thus, A(3) is an element in list A, and 3 is the subscript; H(2, 7) is an element in table H and 2 and 7 are subscripts. The subscript may be a constant, a variable, or a legitimate arithmetic expression and must equal a nonnegative integer value.

The name of a list or table can be any alphabetic character. Thus, any of the letters A, B, C, D, ... Y, Z can be used to name lists and tables. A program, however, cannot contain a list and a table with the same name.

If you want to use subscripts that are greater than 10, you must inform the computer with a DIM (dimension) statement. For example, the statement

$$200 \quad \text{DIM A}(40)$$

informs the computer that your program will need 40 computer storage locations labeled A(1), A(2), A(3), ... A(40). The statement

$$100 \quad \text{DIM X}(25), \text{F}(10, 15)$$

directs the computer to reserve 25 storage locations for a list named X and 10 × 15, or 150 storage locations for a table named F. DIM statements are usually placed at the beginning of a BASIC program.

Consider a list of eight numbers and suppose the name of the list is X. We write $X(1)$ to refer to the first element or number in the list, $X(2)$ to refer to the second element in the list, and $X(8)$ to refer to the last element in the list. We can find the sum of all eight numbers in the list through the statement.

30   LET S = $X(1) + X(2) + X(3) + X(4) + X(5) + X(6) + X(7) + X(8)$.

This statement would accomplish the task, but what if the list contained 200 numbers and we wanted to add all 200? Obviously, we must find a more efficient way. The following program segment uses a program loop to add 200 elements of a list named X. The sum of the elements would now be stored in variable $S$.

```
100   LET S = 0
200   FOR N = 1 TO 200
300   LET S = S + X(N)
400   NEXT N
```

The following BASIC program reads seven test scores for a particular student, averages those test scores, and prints the result.

```
100 REM GRADE AVERAGE
110 REM READ TEST SCORES INTO LIST R
120 FOR I = 1 TO 7
130 READ R(I)
140 NEXT I
```

```
150 REM OBTAIN SUM OF TEST SCORES
160 LET S = 0
170 FOR I = 1 TO 7
180 LET S = S + R(I)
190 NEXT I
200 REM COMPUTE AVERAGE OF TEST
201 REM SCORES
210 LET A = S/7
220 PRINT "AVERAGE GRADE IS"; A
230 DATA 76, 70, 89, 65, 55, 91, 93
240 END

RUN

AVERAGE GRADE IS 77
```

The seven student test scores are stored in a list called R. Pictorally the list may be visualized as follows:

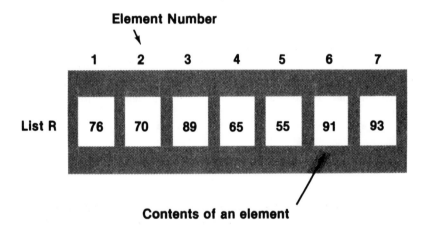

**Contents of an element**

The name R refers to the entire list. Each element of the list can contain a value. These elements are numbered from one to seven. Note that the element number is distinctly different from the contents of that element.

In the program, statements 120 through 140 cause the seven test scores to be stored in list R. Statements 160 through 190 accumulate the sum of the test scores in variable *S*. Statement 210 finds the average by dividing the sum of the test scores by seven. Statement 220 causes the average to be printed.

A table is composed of horizontal rows and vertical columns. To find any element in a table we must identify two indices: one to specify which element in the row (row index), and the other to indicate which column (column index). The indices are appended to the table name within parentheses. Suppose that T is the name of a table with eight rows and eight columns. Then T(2, 6) would refer to the second element in the sixth column.

Before we look at a number summation program, let us consider the procedure for setting all elements of a table to zero by means of a three-statement loop. The following group of statements sets to zero all elements of a 100 by 150 table named S.

```
100 DIM S(100, 150)
110 FOR I = 1 TO 100
120 FOR J = 1 TO 150
130 LET S(I, J) = 0
140 NEXT J
150 NEXT I
```

Given the following table of numbers in six rows and five columns, find the sum of all numbers except those on a border or, in other words, the sum of the numbers enclosed in the rectangle. Print the entire table (named H), the numbers that do not lie on a border, and the computed sum.

| | | | | |
|---|---|---|---|---|
| 40 | 32 | 16 | 19 | 24 |
| 17 | 06 | 31 | 92 | 91 |
| 16 | 22 | 11 | 47 | 55 |
| 83 | 64 | 87 | 71 | 03 |
| 14 | 33 | 16 | 92 | 18 |
| 36 | 14 | 09 | 76 | 84 |

The following program first initializes a variable *S*, which will ultimately be the sum of the nonborder numbers. Statements 120 through 160 cause 30 numbers from the data bank to be read into table H. The sum of the selected numbers is computed by statements 170 through 210. Statements 230 through 290 cause all the numbers in table H to be printed. Statements 320 through 380 cause the selected numbers to be printed, and statement 400 causes a message and the sum to be printed. Statements 410 through 460 are the DATA statements specifying the 30 numbers that were read into the data bank.

```
10    REM SUM OF CENTER ELEMENTS
20    LET S=0
30    FOR W=1 TO 6
40    FOR X=1 TO 5
50    READ H(W,X)
60    NEXT X
70    NEXT W
80    FOR A=2 TO 5
90    FOR B=2 TO 4
100   LET S=S+H(A,B)
110   NEXT B
120   NEXT A
130   REM PRINT ENTIRE TABLE
140   FOR I=1 TO 6
150   FOR K=1 TO 5
160   PRINT H(I,K);
170   NEXT K
180   PRINT
190   PRINT
200   NEXT I
210   REM PRINT THREE BLANK LINES
220   PRINT
230   PRINT
240   PRINT
250   REM PRINT NUMBERS THAT DO NOT LIE ON THE BORDER
260   FOR I=2 TO 5
270   FOR J=2 TO 4
280   PRINT H(I,J);
290   NEXT J
300   PRINT
310   PRINT
320   NEXT I
330   REM PRINT SUM OF NUMBERS
340   PRINT "SUM OF NUMBERS THAT DO NOT LIE"
341   PRINT "ON THE BORDER ="; S
350   DATA 40,32,16,19,24
360   DATA 17,6,31,92,91
370   DATA 16,22,11,47,55
380   DATA 83,64,87,71,3
390   DATA 14,33,16,92,18
```

```
400   DATA 36,14,9,76,84
410   END

RUN

40    32    16    19    24
17     6    31    92    91
16    22    11    47    55
83    64    87    71     3
14    33    16    92    18
36    14     9    76    84

 6    31    92
22    11    47
64    87    71
33    16    92


SUM OF NUMBERS THAT DO NOT LIE
ON THE BORDER = 572
```

## 2.7  ADVANCED FEATURES

Up to this point we have discussed the following statements:

|  |  |  |
|--|--|--|
| READ | LET | REMARK (or REM) |
| DATA | GOTO | GOSUB |
| INPUT | IF/THEN | RETURN |
| PRINT | FOR | DIM |
| END | NEXT | Functions |

Using just these BASIC statements you can write programs to solve many kinds of problems.

The BASIC language, however, includes other statements to aid you in the development of programs. Discussed in this section are the following statements:

STOP
RESTORE
ON/GOTO
TAB

Section 2.8 introduces matrix operations and the MAT statements.

## RESTORE Statement

The BASIC language has a provision for using the same data more than once in a program. If a program had been written and the data inserted, we could send the computer back to the beginning of the data list by using a statement called RESTORE. Once data has been entered, the computer reads the DATA statements in the order of their occurrence. By adding a RESTORE statement, the computer reverts to the first DATA statement in the program. For example, consider the following program:

```
100 PRINT "NUMBERS ARE:";
110 FOR N = 1 TO 4
120 READ X
130 PRINT X;
140 NEXT N
150 RESTORE
160 PRINT
170 PRINT
180 READ W, X, Y, Z
190 PRINT "SUM OF NUMBERS:";
200 PRINT W+X+Y+Z
210 DATA 34, 98
220 DATA 56, 88
230 END

RUN

NUMBERS ARE: 34  98  56  88
SUM OF NUMBERS: 276
```

The computer reads and prints the four numbers 34, 98, 56, and 88, and restores the data so that it can be used again by statement 180. This statement is particularly useful when the same data are used at several places in the program.

## STOP Statement

The STOP statement is used to terminate the computation at any point in a BASIC program. It is equivalent to a GOTO statement that directs program control to the END statement. The statement consists of a line number followed by the keyword STOP. It differs from the END statement in that it signifies the logical termination of the program, not the physical termination of the program.

The STOP statement can be inserted anywhere in the program (except at the very end) where you might wish the computer to halt execution.

## ON/GOTO Statement

The ON/GOTO statement permits transfer of control to one of a group of statements, with the particular one chosen during the run on the basis of results computed in the execution of the program. The statement is of the form

$$ln \quad \text{ON expression GOTO } (\ln_1), (\ln_2), (\ln_3), \ldots$$

where the expression is any valid BASIC expression (other than a string or string variable), and the subscripts on the bracketed line numbers of statements in the program indicate their sequence in the GOTO statement. Execution of this statement causes statement $(\ln_i)$ to be executed next, where $i$ is the integer value of the expression. If computation of the expression produces a result other than an integer, the result is truncated to its integer value, regardless of the value of the fractional part. For example, if the expression is $A + B$ and $A = 1.7$ and $B = 3.2$,

$$300 \quad \text{ON A + B GOTO 100, 210, 120, 130, 440}$$

is executed, control is transferred to statement number 130, the fourth line number in the GOTO statement.

The user must be sure that the expression in the ON/GOTO statement produces a result of at least 1 and no more than the number of line number labels contained in the statement. For example, the expression in the statement

$$200 \quad \text{ON X} - \text{Y} + 2 \text{ GOTO 300, 160, 410, 280}$$

must produce an integer value in the range 1 to 4.

**TAB Function**

The TAB function is used for spacing. It is used in a PRINT statement, and TAB($N$) refers to the $N$th column of a print line or a display line. The execution of the statement PRINT TAB(10) would cause the cursor or printing element to move to a position 10 spaces from the left, i.e. column 10. This feature allows one to space printed output on a line in any position. For example,

$$100 \quad \text{LET X} = 98.30$$
$$200 \quad \text{PRINT TAB(5); X}$$

will result in:

$$98.30$$

with the 9 in column 6. On the other hand,

$$100 \quad \text{LET X} = 98.30$$
$$200 \quad \text{PRINT X; TAB(14); X}$$

will result in:

$$98.30 \qquad 98.30$$

with the first 9 being printed in column 2 and the second 9 in column 15.

The following program uses the TAB function to print a diagonal line across a page.

```
100 REM DIAGONAL LINE
110 FOR K = 1 TO 10
120 PRINT TAB(K); "PASCAL"
130 NEXT K
140 END

RUN

    PASCAL
     PASCAL
      PASCAL
       PASCAL
        PASCAL
         PASCAL
          PASCAL
           PASCAL
            PASCAL
             PASCAL
```

Let us now look at a program that produces a table with column headings starting in columns 11, 23, and 27.

```
100 REM POWER TABLE
110 PRINT TAB(10); "X"; TAB(22); "X SQ"; TAB(36); "X CU"
120 FOR X = 1 TO 15
130 PRINT TAB(10); X; TAB(22); X*X; TAB(36); X*X*X
140 NEXT X
150 END
```

| X | X SQ | X CU |
|---|------|------|
| 1 | 1 | 1 |
| 2 | 4 | 8 |
| 3 | 9 | 27 |
| 4 | 16 | 64 |
| 5 | 25 | 125 |
| 6 | 36 | 216 |
| 7 | 49 | 343 |
| 8 | 64 | 512 |
| 9 | 81 | 729 |
| 10 | 100 | 1000 |
| 11 | 121 | 1331 |
| 12 | 144 | 1728 |
| 13 | 169 | 2197 |
| 14 | 196 | 2744 |
| 15 | 225 | 3375 |

Note that TAB functions are used in both statement 110 and statement 130 so that the heading for the table and the numbers they label are in the correct columns.

## 2.8  MATRICES AND MATRIX OPERATIONS

In Section 2.6, the reader was introduced to arrays of numbers; or, as they are commonly called in BASIC, lists and tables. An array of numbers can also be called a matrix. In this section, we will refer to two-dimensional arrays as matrices and we shall also learn how to perform many important operations of matrix algebra. We will cover some of the fundamental points about matrices in order that the eleven BASIC matrix operation statements can be discussed.

The following arrays of numbers:

$$\begin{bmatrix} 1 & 6 & 2 \\ 3 & 1 & 9 \end{bmatrix} \quad \begin{bmatrix} 3 & 1 \\ 4 & 2 \end{bmatrix} \quad \begin{bmatrix} 1 & 2 & 2 \\ 6 & 4 & 5 \\ 3 & 1 & 6 \end{bmatrix}$$

are matrices. A matrix, then, is a rectangular or square array of numbers, arranged in rows and columns. The first example, with two rows and three columns, is called a 2 × 3 matrix (read "2 by 3"), the second example is a 2 × 2 matrix, and the third example is a 3 × 3 matrix. The second and third examples are called square matrices since they have the same number of rows as columns.

In this section, matrices will be written with brackets around the rectangular array of numbers. Other notations are acceptable and are used in other publications. For example, parentheses

$$\begin{pmatrix} 12 & 14 \\ 26 & 32 \\ 41 & 16 \end{pmatrix}$$

and double-bar notations

$$\begin{Vmatrix} 63 & 12 & 37 \\ 41 & 61 & 64 \\ 18 & 19 & 23 \end{Vmatrix}$$

are sometimes used.

Two matrices are equal if, and only if, they are of the same order, and each entry of one is equal to the corresponding entry of the order. For example,

$$\begin{bmatrix} a_{11} & a_{12} \\ a_{21} & a_{22} \end{bmatrix} = \begin{bmatrix} b_{11} & b_{12} \\ b_{21} & b_{22} \end{bmatrix}$$

if and only if $a_{11} = b_{11}$, $a_{21} = b_{21}$, $a_{12} = b_{12}$, and $a_{22} = b_{22}$.

The BASIC language contains eleven statements designed for matrix operations. Each begins with MAT, followed by the specific operation. Table 1 lists examples of the eleven statements, along with their identifications. The matrix in each case is named C, and $A$ and $B$ represent values involved in the given operation. All other characters are fixed parts of the statements.

Before the BASIC matrix operations can be used, each matrix must be declared in a DIM statement, which reserves computer storage for it. For example, the statement

$$10 \quad \text{DIM A(2, 4), B(10, 12), C(30, 30)}$$

reserves storage for a 2 × 4 matrix named A, a 10 × 12 matrix named B, and a 900-element square matrix named C. The first number in parentheses specifies the number of rows in the matrix; the second number specifies the number of columns. Thus a 4 × 6 matrix could be dimensioned by the statement 10   DIM P(4, 6), which specifies that the matrix named P has four rows and six columns. Note that a matrix name must be a single letter, such as D, R, or W.

Before any computations can be made by means of a matrix statement, the size of the matrix must be established and values must be assigned to it. The size of a matrix named C may be established by any of the following statements: DIM C(M, N), MAT READ C(M, N), MAT C = ZER(M, N), MAT C = IDN(M, N), or MAT C = CON(M, N), where $M$ is the number of rows in the matrix and $N$ is the number of columns.

Values may be assigned to the matrix C by any of the following statements: MAT READ C, MAT C = ZER, MAT C = IDN, or MAT C = CON. The

**Table 2.1**  Matrix Operation Statements Used in BASIC

| Statement | Identification |
|---|---|
| MAT READ C | Read Matrix |
| MAT PRINT C | Print Matrix |
| MAT C = TRN(A) | Transpose Matrix |
| MAT C = ZER | Zero Matrix |
| MAT C = IDN | Identity Matrix |
| MAT C = CON | J Matrix |
| MAT C = A + B | Add Matrix |
| MAT C = A − B | Subtract Matrix |
| MAT C = (A)·B | Scalar Multiplication |
| MAT C = A·B | Multiply Matrix |
| MAT C = INV(A) | Invert Matrix |

MAT READ statement is used to read values from a DATA statement into a specified matrix. The general form is

MAT READ v

where v is the name of a matrix that has been dimensioned by a DIM statement. The MAT READ statement can also specify more than one previously dimensioned matrix. The general form is then

MAT READ $v_1, v_2, \ldots, v_n$

The sequence of statements

```
10   DIM A(4, 3)
20   MAT READ A
30   DATA 1, 2, 3, 4, 5, 6, 7, 8, 9, 10, 11, 12
```

would cause the values 1, 2, and 3 to be established as the values for row 1 of the matrix named A; 4, 5, and 6 for row 2; 7, 8, and 9 for row 3; and 10, 11, and 12 for row 4. Note that the MAT READ statement reads the values from the DATA statements in row order, i.e., row 1 first, then row 2, then row 3, then row 4, etc.

It is possible to reserve more storage than is actually needed. When in doubt, indicate a larger dimension than you expect to use. This precaution will allow you to expand the size of the matrix at some future date without changing the DIM statement, provided that you do not exceed the size that was originally dimensioned. For example, if you wanted to create a list of twelve numbers, you might write the program segment shown below. Statements 300 and 700 could have been eliminated by writing statement 400   FOR J = 1 to 12, but the form as written allows for lengthening list X by changing only statement 700, as long as the length of the list does not exceed 100.

```
100   REM  MATRIX EXAMPLE
200   DIM X(100)
300   READ K
400   FOR J = 1 TO K
500   READ X(J)
600   NEXT J
700   DATA  12
800   DATA  6,7,4,3,10,14,36,9,12,9,1,17
```

The general form of the MAT PRINT statement is

MAT PRINT v

where v is the name of a matrix that has been dimensioned in a DIM statement. Like the MAT READ statement, the MAT PRINT statement can contain the names of several previously dimensioned matrices.

There are two types of printing formats, regular and packed. If the matrix name is followed by a semicolon, the matrix is printed in packed format, otherwise, regular format is used. The program shown below establishes values for and prints the matrix W in packed format.

```
100 MATRIX EXAMPLE
110 DIM W(2,5)
120 MAT READ W
130 MAT PRINT W;
140 DATA 4,24,8,36,12,3,18,6,27,9
150 END

RUN

4       24      8       36      12
3       18      6       27      9
```

A matrix is transposed when the elements in the rows become the elements in the column, and vice versa. For example, the 2 × 4 matrix

$$\begin{bmatrix} 8 & 6 & 7 & 2 \\ 3 & 5 & 6 & 1 \end{bmatrix}$$

when transposed becomes the 4 × 2 matrix

$$\begin{bmatrix} 8 & 3 \\ 6 & 5 \\ 7 & 6 \\ 2 & 1 \end{bmatrix}$$

The general form of the BASIC matrix transposition statement is

$$\text{MAT } v_1 = \text{TRN}(v_2)$$

where $v_1$ and $v_2$ are matrices. It would transpose matrix $v_1$ and assign it in the new form to matrix $v_2$. For example, the program shown below uses the matrix transposition statement to transpose a $4 \times 5$ matrix. The program assigns the original values to matrix S, transposes matrix S, assigns the transposed values to matrix Y, and prints both matrices.

```
100 DIM S(4,5),Y(5,4)
110 MAT READ S
120 PRINT "MATRIX S"
130 PRINT
140 MAT PRINT S
150 MAT Y = TRN(S)
160 PRINT "TRANSPOSE OF MATRIX S IS"
170 PRINT
180 MAT PRINT Y
190 DATA 6,14,3,5,24,2,6,15,9,13,12,2,5,8,11,1,23,16,4,9
200 END

RUN

MATRIX S

6        14       3        5        24
2        6        15       9        13
12       2        5        8        11
1        23       16       4        9

TRANSPOSE OF MATRIX S IS

6        2        12       1
14       6        2        23
3        15       5        16
5        9        8        4
24       13       11       9
```

A matrix whose elements are all zeros is called the zero matrix. The BASIC statement

$$\text{MAT } v = \text{ZER}$$

can be used to place zeros in all elements of the matrix represented by v. For example, the statement

$$800 \quad \text{MAT R} = \text{ZER}$$

will cause all elements of matrix R to be replaced by zeros.

The identity (or unit) matrix is a square matrix with all elements zeros except those that are on its main diagonal, which are all ones. The BASIC statement

$$\text{MAT v} = \text{IDN}$$

may be used to set up matrix v as an identity matrix. Note that only a square matrix can be set up as an identity matrix.

A matrix whose elements are all ones is called the J matrix. The statement

$$\text{MAT v} = \text{CON}$$

may be used to place ones in all elements of matrix v.

Addition of two matrices of the same order is accomplished by adding the elements of one to the corresponding elements of the other, and placing the sums in a third matrix of the same dimension. The addition of matrices a and b to produce matrix c can be expressed by the equation

$$c_y = a_y + b_y$$

where $c_y$, $a_y$, and $b_y$ are corresponding elements of the three matrices. The general form of the matrix addition statement is

$$\text{MAT v}_1 = v_2 + v_3$$

To illustrate this instruction, let us consider a BASIC program to perform the addition

$$\begin{bmatrix} 2 & 1 & 7 \\ 4 & -6 & 2 \\ 0 & 3 & -3 \end{bmatrix} + \begin{bmatrix} 16 & 8 & -1 \\ 4 & 6 & 9 \\ -3 & 2 & 8 \end{bmatrix} = \begin{bmatrix} ? \end{bmatrix}$$

and print the result. The program shown below assigns the name X to the first matrix, Y to the second, and Z to the summation matrix.

```
100 DIM X(3,3),Y(3,3),Z(3,3)
110 MAT READ X,Y
120 PRINT "MATRIX X"
130 PRINT
140 MAT PRINT X
150 PRINT
160 PRINT "MATRIX Y"
170 PRINT
180 MAT PRINT Y
190 PRINT
200 REM ADD MATRICES X AND Y
210 MAT Z = X+Y
220 PRINT "SUM OF MATRICES X AND Y IS LOCATED IN MATRIX Z"
230 PRINT
240 PRINT "MATRIX Z"
250 PRINT
260 MAT PRINT Z
270 DATA 2,1,7,4,-6,2,0,3,-3,16,8,-1,4,6,9,-3,2,8
280 END

RUN

MATRIX X

 2          1          7

 4         -6          2

 0          3         -3
MATRIX Y

 16         8         -1

 4          6          9

-3          2          8
SUM OF MATRICES X AND Y IS LOCATED IN MATRIX Z
MATRIX Z

 18         9          6

 8          0          11

-3          5          5
```

Subtraction of two matrices of the same order is accomplished by the same method that is used for addition, except that the corresponding elements are, of course, subtracted. The BASIC statement is

$$\text{MAT } v_1 = v_2 - v_3$$

To illustrate, let us perform the subtraction

$$\begin{bmatrix} 6 & 1 & 6 \\ 4 & 0 & 7 \\ 2 & 3 & 2 \end{bmatrix} - \begin{bmatrix} 2 & 4 & 6 \\ 6 & 3 & 2 \\ 1 & 7 & 4 \end{bmatrix} = \begin{bmatrix} ? \end{bmatrix}$$

The program shown below names the minuend matrix A, the subtrahend matrix B, and the difference matrix C.

```
100 DIM C(3,3) ,A(3,3) ,B(3,3)
110 MAT READ A,B
120 MAT C=A-B
130 PRINT "MATRIX A"
140 MAT PRINT A
150 PRINT
160 PRINT "MATRIX B"
170 MAT PRINT B
180 PRINT
185 PRINT
190 PRINT "MATRIX C"
200 MAT PRINT C
210 DATA 6,1,6,4,0,7,2,3,2,2,4,6,6,3,2,1,7,4
220 END

RUN

MATRIX A

   6           1           6
   4           0           7
   2           3           2
```

MATRIX B

| 2 | 4 | 6 |
|---|---|---|
| 6 | 3 | 2 |
| 1 | 7 | 4 |

MATRIX C

| 4 | -3 | 0 |
|---|----|---|
| -2 | -3 | 5 |
| 1 | -4 | -2 |

Scalar multiplication is performed by multiplying each element of a matrix by the same factor, called a scalar. In the following example, the scalar is 6:

$$6 \times \begin{bmatrix} 6 & 4 \\ 3 & 2 \\ 1 & 7 \end{bmatrix} = \begin{bmatrix} 36 & 24 \\ 18 & 12 \\ 6 & 42 \end{bmatrix}$$

In BASIC, the scalar may be a constant, a variable, or an expression. The general form of the statement is

$$\text{MAT } v_1 = (s) * v_2$$

Note that parentheses around the scalar, $s$, distinguish this form of multiplication from a matrix multiplication. The program shown below gives the matrix

$$\begin{bmatrix} 6 & 1 & 2 \\ 2 & 4 & 4 \\ 3 & 7 & 5 \end{bmatrix}$$

the name W and multiplies it first by the constant 6 and then by the expression $N \times 13$, in which $N$ is equal to 8 as indicated in statement 230. The program causes both the original matrix and the two calculated matrices to be printed.

```
100 REM MATRIX EXAMPLE
110 DIM W(3,3),S(3,3)
120 MAT READ W
130 PRINT "MATRIX W"
140 MAT PRINT W
150 PRINT
160 REM -MULTIPLY MATRIX W BY 6
170 MAT S = (6)*W
180 PRINT
190 PRINT "MATRIX W MULTIPLIED BY 6"
200 MAT PRINT S
210 PRINT
220 REM ---MULTIPLY MATRIX W BY N*13
230 LET N = 8
240 MAT S = (N*13)*W
250 PRINT
260 PRINT "MATRIX W MULTIPLIED BY N*13"
270 MAT PRINT S
280 DATA 6,1,2,2,4,4,3,7,5
290 END

RUN

MATRIX W

6          1          2

2          4          4

3          7          5

MATRIX W MULTIPLIED BY 6

36         6          12

12         24         24

18         42         30

MATRIX W MULTIPLIED BY N*13

624        104        208

208        416        416

312        728        520
```

The multiplication of two matrices is an interesting operation. The products of the row elements of one, and the corresponding column elements of the other, are added together to form the elements of a product matrix having the same number of rows as the first, and the same number of columns as the second. Consequently, if two matrices are to be multiplied, the first must have the same number of elements in each row as the second has in each column. To illustrate, let us compute the product of the following two matrices:

$$\begin{bmatrix} a_1 & a_2 & a_3 \\ b_1 & b_2 & b_3 \end{bmatrix} \times \begin{bmatrix} x_1 & x_2 \\ y_1 & y_2 \\ z_1 & z_2 \end{bmatrix}$$

The clearest way to explain this operation is to show how the individual products are obtained and combined to form the product matrix:

$$\begin{bmatrix} a_1x_1 + a_2y_1 + a_3z_1 & a_1x_2 + a_2y_2 + a_3z_2 \\ b_1x_1 + b_2y_1 + b_3z_1 & b_1x_2 + b_2y_2 + b_3z_2 \end{bmatrix}$$

If we now assign arbitrary values to our two original matrices, we can compute their products as follows:

$$\begin{bmatrix} 1 & 3 & 2 \\ 0 & -1 & 2 \end{bmatrix} \times \begin{bmatrix} 3 & 6 \\ -1 & 2 \\ 2 & 1 \end{bmatrix} =$$

$$\begin{bmatrix} 1(3) - 3(-1) + 2(2) & 1(6) + 3(2) + 2(1) \\ 0(3) + (-1)(-1) + 2(2) & 0(6) + (-1)(2) + 2(1) \end{bmatrix}$$

$$= \begin{bmatrix} 4 & 14 \\ 5 & 0 \end{bmatrix}$$

The BASIC statement for multiplying two matrices is

$$\text{MAT } v_1 = v_2 * v_3$$

The program shown below determines the product of the following two matrices:

$$\begin{bmatrix} 1 & -2 & -3 \\ 2 & 3 & -4 \\ 5 & 0 & 2 \end{bmatrix} \times \begin{bmatrix} 2 & 1 & -1 \\ 4 & 3 & 2 \\ 0 & 1 & -1 \end{bmatrix}$$

```
100 REM MATRIX EXAMPLE
110 DIM C(3,3),A(3,3),B(3,3)
120 MAT READ A,B
130 PRINT "MATRIX A"
140 MAT PRINT A
150 PRINT
160 PRINT "MATRIX B"
170 MAT PRINT B
180 PRINT
190 REM  ---DETERMINE PRODUCT OF A*B
200 MAT C = A*B
210 PRINT "MATRIX C"
220 MAT PRINT C
230 DATA 1,-2,-3,2,3,-4,5,0,2,2,1,-1,4,3,2,0,1,-1
240 END

RUN

MATRIX A

1          -2          -3

2           3          -4

5           0           2

MATRIX B

2           1          -1

4           3           2

0           1          -1
```

MATRIX C

| 16 | 7 | 8 |
|----|---|-----|
| 10 | 7 | -7 |
| 18 | 7 | -7 |

Previously, we encountered the identity matrix in which the elements of the main diagonal are ones and the other elements are zeros. Let us now see the reason for that name by multiplying a square matrix by the identity matrix:

$$\begin{bmatrix} a_1 & a_2 & a_3 \\ b_1 & b_2 & b_3 \\ c_1 & c_2 & c_3 \end{bmatrix} \times \begin{bmatrix} 1 & 0 & 0 \\ 0 & 1 & 0 \\ 0 & 0 & 1 \end{bmatrix} = \begin{bmatrix} a_1 & a_2 & a_3 \\ b_1 & b_2 & b_3 \\ c_1 & c_2 & c_3 \end{bmatrix}$$

The principle involved here may, of course, be applied to a square matrix of any size. If we represent the identity matrix by the symbol i, then a $\times$ i = a for any matrix a.

For every nonsingular square matrix there exists an inverse; the product of the matrix and its inverse is the identity matrix. Thus

$$a \times a' = a' \times a = i$$

where a is a nonsingular square matrix and a' is its inverse. Let us now determine the inverse, a', of a given a by means of the equation a $\times$ a' = i:

$$\underset{a}{\begin{bmatrix} 9 & 5 \\ 7 & 4 \end{bmatrix}} \times \underset{a'}{\begin{bmatrix} x & u \\ y & v \end{bmatrix}} = \underset{i}{\begin{bmatrix} 1 & 0 \\ 0 & 1 \end{bmatrix}}$$

By the rules of matrix multiplication, we obtain the following equations:

$$9x + 5y = 1 \qquad 9u + 5v = 0$$
$$7x + 4y = 0 \qquad 7u + 4v = 1$$

Solving these equations, we get $x = 4$, $y = -7$, $u = -5$, and $v = 9$. Therefore, the inverse (a') is

$$\begin{bmatrix} 4 & -5 \\ -7 & 9 \end{bmatrix}$$

By the same procedure, the inverse of the matrix

$$\begin{bmatrix} 1 & 2 & 0 \\ -1 & 1 & 3 \\ 0 & 1 & -1 \end{bmatrix}$$

is

$$\begin{bmatrix} 4/6 & -2/6 & -1 \\ 1/6 & 1/6 & 3/6 \\ 1/6 & 1/6 & -3/6 \end{bmatrix}$$

The answer can be checked by performing the following matrix multiplication:

$$\begin{bmatrix} 4/6 & -2/6 & -1 \\ 1/6 & 1/6 & 3/6 \\ 1/6 & 1/6 & -3/6 \end{bmatrix} \times \begin{bmatrix} 1 & 2 & 0 \\ -1 & 1 & 3 \\ 0 & 1 & -1 \end{bmatrix} = \begin{bmatrix} 1 & 0 & 0 \\ 0 & 1 & 0 \\ 0 & 0 & 1 \end{bmatrix}$$

Finding the inverse of a large matrix is obviously a complex procedure, but in scientific work it is often necessary to find the inverse of a matrix with several hundred rows. Therefore, several different methods have been developed for performing this computation. In BASIC, the inverse of a matrix may be determined by using the statement

$$\text{MAT } v = \text{INV}(v_2)$$

Let us now consider the program shown below, which determines the inverse of matrix R. This program prints the original matrix R (statement 150), computes its inverse J (statement 180), and prints J (statement 220).

```
100 REM MATRIX EXAMPLE
110 DIM R(4,4),J(4,4)
120 MAT READ R
130 PRINT "MATRIX R"
140 PRINT
150 MAT PRINT R
```

```
160 PRINT
170 REM  -DETERMINE INVERSE OF MATRIX R
180 MAT J = INV(R)
190 PRINT
200 PRINT "INVERSE OF MATRIX R"
210 PRINT
220 MAT PRINT J
230 DATA 2,3,8,4,7,1,6,0,8,0,3,3,5,2,0,4
240 END

RUN

MATRIX R
```

| 2 | 3 | 8 | 4 |
| 7 | 1 | 6 | 0 |
| 8 | 0 | 3 | 3 |
| 5 | 2 | 0 | 4 |

```
INVERSE OF MATRIX R
```

| -8.02675E-2 | .100334 | .013378 | 7.02342E-2 |
| -3.00999E-2 | .287625 | -.494982 | .401338 |
| 9.86621E-2 | 1.67298E-3 | 6.68892E-2 | -.148829 |
| .115385 | -.269231 | .23077 | -3.84618E-2 |

An important application of matrix inversion in BASIC is the solution of simultaneous linear equations such as the following:

$$x + 2y + 3z = 26$$
$$3x + 5y + 2x = 39$$
$$2x + 4y + x = 27$$

It is possible to represent these equations as a simple matrix equation in the following manner:

$$\begin{bmatrix} 1 & 2 & 3 \\ 3 & 5 & 2 \\ 2 & 4 & 1 \end{bmatrix} \times \begin{bmatrix} x \\ y \\ z \end{bmatrix} = \begin{bmatrix} 26 \\ 39 \\ 27 \end{bmatrix}$$

To determine the inverse of the square matrix in our equation, we will name it M, and then write and execute the program shown below.

```
100 DIM M(3,3),D(3,3)
200 MAT READ M
300 MAT D = INV(M)
400 MAT PRINT D
500 DATA 1,2,3,3,5,2,2,4,1
600 END

RUN

-.6           2           -2.2

 .2          -1            1.4

 .4           0            -.2
```

With the inverse printed by this program, we can now determine the values of x, y, and z. We multiply the inverse matrix by

$$\begin{bmatrix} 26 \\ 39 \\ 27 \end{bmatrix}$$

since this would give us

$$\begin{bmatrix} 1 & 0 & 0 \\ 0 & 1 & 0 \\ 0 & 0 & 1 \end{bmatrix} \times \begin{bmatrix} x \\ y \\ z \end{bmatrix} = \begin{bmatrix} -.6 & 2 & -2.2 \\ .2 & -1 & 1.4 \\ .4 & 0 & -.2 \end{bmatrix} \times \begin{bmatrix} 26 \\ 39 \\ 27 \end{bmatrix}$$

A BASIC program to perform this multiplication is given by the program shown below. The matrix printed by this program gives us the unknown values for the original equations: $x = 3$, $y = 4$, and $z = 5$.

```
10 DIM D(3,3), S(3,1), A(3,1)
20 MAT READ D, S
30 MAT A = D * S
40 MAT PRINT A
50 DATA -.6, 2, -2.2, .2, -1, 1.4, .4, 0, -.2
60 DATA 26, 39, 27
70 END

RUN

   3.

   4.

   5.
```

## 2.9  SAMPLE BASIC PROGRAMS

The beginning computer user, whether he or she is a college student, a professional, a scientist, a teacher, a businessperson, a secondary school student, or anyone else, is advised to start with problems that are relatively easy to understand. With a clear understanding of simple problems he or she may gradually move on to more complicated problems. The reader should remember that there are often many possible approaches to any given problem.

The sample problems given here are written in a straightforward manner and contain comment statements to help the reader understand each program.

### An Island of Money

King Bigfoot, leader of a wealthy South Seas island, was distributing his wealth to the one million natives who lived on the island. He gave $1 to the first native. The next two natives got $2 each. The next three natives each got $3, and so on. The BASIC program shown below determines what the millionth native received.

```
100 REM ISLAND OF MONEY
200 LET A = 0
300 LET B = 1
400 LET A = A + B
500 IF A >= 1000000 THEN 800
600 LET B = B + 1
700 GOTO 400
800 PRINT "THE MILLIONTH NATIVE GOT $"; B
900 END

RUN


THE MILLIONTH NATIVE GOT $ 1414
```

## Number Sum

The BASIC program shown below determines the sum of all numbers between 91 and 989 that are divisible by 13.

```
10 REM NUMBER SUM
20 LET S = 0
30 FOR N = 91 TO 989 STEP 13
40 LET S = S + N
50 NEXT N
60 PRINT "NUMBER SUM="; S
70 END

RUN

NUMBER SUM= 37765
```

## Coin Tossing

The BASIC program shown below uses the RND (random number) library function to simulate heads and tails (heads if the random number is less than 0.5; tails if the random number is 0.5 or greater). The program simulates the flipping of 20 coins and prints out the number of heads and tails obtained.

```
100 REM COIN TOSSING
110 LET T = 0
120 LET H = 0
130 FOR N = 1 TO 20
140 LET R = RND(2)
150 IF R <= 1 THEN 190
160 PRINT "TAIL"
170 LET T = T + 1
180 GOTO 210
190 PRINT "HEAD"
200 LET H = H + 1
210 NEXT N
220 PRINT "NUMBER OF TAILS ="; T
230 PRINT "NUMBER OF HEADS ="; H
240 END
```

## Saving Money

Mary Thriftway opens a savings account at the First International Bank. Each day she deposits twice the amount of the previous day. She puts one dollar in the bank on day 1, two dollars on day 2, four dollars on day 3, etc. The following program determines the amount Mary has in the bank at the end of 18 days.

```
100 REM MARY'S SAVINGS ACCOUNT
110 PRINT "DEPOSIT", AMOUNT", "TOTAL AMOUNT"
120 PRINT "DATE", "OF DEPOSIT", "IN BANK"
130 PRINT
140 LET T=0
150 FOR N=0 TO 17
160 REM AMOUNT OF DEPOSIT - S
170 LET S=2↑N
180 REM TOTAL AMOUNT IN BANK - T
190 LET T=T + S
200 PRINT N+1, S, T
210 NEXT N
220 END
```

## 2.10 SUMMARY

The BASIC programming language can be used by almost anyone to solve many problems. The language includes input, mathematics, control, and output instructions. Input statements are READ, DATA, and INPUT. The output statement is the PRINT statement.

The BASIC arithmetic instruction is very similar to algebraic notation, using the following arithmetic operators: + (addition), − (subtraction), * (multiplication), / (division), and ↑ (exponentiation). The equal sign (=) is different from the algebraic sign and in BASIC means "replaced by." For example, calculate the value of the expression to the right of the "replaced by" symbol and store that value in the storage area named by the variable to the left of the symbol. Parentheses are used in the same way as in algebra.

BASIC control commands include unconditional and conditional branching statements as well as a pair of looping statements. The unconditional branch is the GOTO statement, which causes the computer to branch to the statement specified by the line number included in the GOTO command. The conditional branch is the IF/THEN statement, which causes the computer to branch depending upon whether the condition tests true or false. The condition tests discussed in this chapter include greater than (>), greater than or equal to (> =), less than (<), less than or equal to (< =), and not equal (< >). The FOR/NEXT statement combination provides a convenient method for program looping.

The END statement is used to terminate every BASIC program. Comments can be added to a BASIC program by using the REM statement.

BASIC includes several functions to perform operations such as square root, absolute value, sine, or cosine. An advanced form of the language also includes a set of matrix statements that allows matrix operations, such as adding two matrices or inverting a matrix.

With the commands discussed in this chapter, you have a good introduction to the BASIC language. There are several other statements and concepts you can learn, of course, but with the information you have and some practice, you can easily master the BASIC language.

## Review Exercises

1. What are some advantages to using a programming language such as BASIC?

2. What types of problems can best be solved in BASIC?

3.  Which of the following are invalid variables in BASIC? Why?

    (a)  X        (c)  A3        (e)  XY3
    (b)  7A       (d)  Z*        (f)  AY

4.  What is the purpose of the REM statement? Indicate the meaning of the following symbols.

    (a)  *        (d)  (
    (b)  ↑        (e)  +
    (c)  /        (f)  −

5.  Represent the following constants in E notation.

    (a)  6 321 420
    (b)  .000 000 023

6.  What is the function of the LET statement?

7.  Convert the following mathematical expressions into BASIC notation.

    (a)  $a^2 + b - 39$
    (b)  $A + b/c$

8.  What does the BASIC statement "20  LET X = X + 1" accomplish?

9.  Write BASIC statements to represent the following algebraic statements.

    (a)  $r = a - 67 + b * 32p$
    (b)  $x = 3x^2 + 4x - 27$

10. Write a BASIC statement to print the message:
    MICROCOMPUTERS ARE SMALL COMPUTERS

11. Write the PRINT statement to cause printing of your full name followed by the data field X.

12. What is the purpose of the comma in a PRINT statement that outputs numerical data? Of a semicolon?

13. Does the END statement have to be the last one in the BASIC program? Does it have to have the largest line number?

14. Write a program to add 631, 4210, 1167, and 36.4.

15. Have the computer compute and print a decimal value for 6/7.

16. Write a program to compute the square and cube values of the first 30 integers. The program should produce a printout in the following form.

| N | Square | Cube |
|---|--------|------|
| 1 | 1 | 1 |
| 2 | 4 | 8 |
| 3 | 9 | 27 |
| 4 | 16 | 64 |
| 5 | 25 | 125 |
| . | . | . |
| . | . | . |
| . | . | . |

17. What is the basic purpose of the GOTO statement?

18. Write a statement that will transfer program control to line number 100.

19. List the general form of the IF statement.

20. In the following program, what is the final value of S?

```
100 LET S = 3
200 FOR J = 1 TO 4
300 IF S < J THEN 500
400 LET S = S + 1
500 NEXT J
600 END
```

21. Write a BASIC program to determine whether Y is between $-30$ and $+30$. If Y falls within these limits, print out TRUE; if not, FALSE.

22. Write a BASIC program to sum the numbers less than 100 that are divisible by 8.

23. Why are READ and DATA statements always linked together?

24. How many values will read by the following statement? 200   READ A, B, C, X1, X2, X3

25. Write a program that will read values for X, Y, and Z and print them, first in the order read and then in reverse order.

26. How does the INPUT statement function? What terminates the INPUT operation?

27. Show two different ways to calculate a square root.

**28.** Nancy Wilson sells bibles at $3.00 each plus $.65 for postage and handling. Write a BASIC program to calculate her total receipts for two weeks during which she sold 158 bibles.

**29.** The Western Freight Company charges the following rates on merchandise shipped from New York to Phoenix.

$70 per ton for the first 12 tons
$40 per ton for every ton over 12

Write a BASIC program to determine how much it would cost to send shipments weighing 14 tons, 42 tons, 6 tons, 130 tons, 2360 tons.

**30.** The student population at Western University increased by 9 percent every year. If the current student population is 2400, how many students can this college expect in 10 years? Write a program to determine the answer.

**31.** Write a BASIC program to determine how many ways change can be made for 50 cents using quarters, dimes, nickels, and pennies.

**32.** Write a BASIC program to compute a table of amounts that an investment of $100 will be at the end of 10, 15, 20, and 25 years at 5%, 5½%, 6%, 6½%, and 7% per year compounded monthly. The program should print the years across the top and the rates in the first column of each row.

**33.** Simple interest is paid on $300 invested at r% for n years. Write a BASIC program to print an interest table for values of r from 0.05% to 6.5% and for integral values of n between 1 and 25.

**34.** You have $100. If you invest it at 6% interest compounded quarterly, how many years will it take before you have $50 000? Write a BASIC program to determine the answer.

**35.** A depositor banks $20 per month. Interest is 7.25% per year compounded monthly. Write a BASIC program to compute the amount the depositor has in her account after 15 years.

**36.** What is an array? A list? A table?

**37.** What is meant by a one-dimensional array? By a two-dimensional array?

**38.** What is a subscripted variable? A subscript?

**39.** If N = 23, to which element in a list of 40 numbers does A(N) refer?

**40.** What is the purpose of the DIM statement? When is it used?

**41.** Write DIM statements for the following.

    (a)   a list of 75 numbers.

    (b)   a table with 20 rows and 14 columns.

    (c)   a 7 by 7 magic square.

    (d)   four lists, A, B, C, and Z, each having 26 items.

**42.** Write a program to read the following numbers into a list named A: 31, 64, 104, 37, 82, 79, 101, 94, 78, 63, 17, 88. The program should print the list in the order given, then in reverse order.

**43.** Write a program to read the numbers 15, 63, 42, 87, 65, 99, 18 into list X, and the numbers 84, 63, 44, 19, 98, 15, 87 into list Y. The program should form and print a new list that contains only those numbers that are in both lists.

**44.** Write a program to print every fourth entry in a list named z, starting with the eighth entry.

**45.** Write a BASIC program to add the corresponding elements of two tables, X and Y. Each table contains 5 rows and 6 columns.

**46.** Write a BASIC program to fill a table named R with the values of the multiplication table up to 12 by 12. The program should print the last two rows.

**47.** Write a BASIC program to fill a table of four rows and five columns with the integers 1, 2, 3, 4, 5, ... 20. Print the table.

# Chapter 3

# PRIME NUMBERS

### Preview

The prime numbers are useful in analyzing problems concerning divisibility, and are interesting because of some of the special properties that they possess as a class. These properties have fascinated mathematicians and others since ancient times, and the richness and beauty of the results of research in this field have been astonishing. Mathematicians have spent enormous amounts of time computing large prime numbers and testing whether certain large numbers are prime. It is in this area that the computer has made a spectacular contribution to number theory (Figure 3.1).

After you finish this chapter, you should be able to:

1. Identify prime numbers, Mersenne primes, twin primes, and composite numbers.
2. See how a computer can be used to generate prime numbers, twin primes, and Mersenne primes.
3. Generate primes using the sieve of Eratosthenes.
4. Write BASIC programs to generate primes.

## 3.1 THE SEQUENCE OF PRIMES

The number 6 is equal to 2 times 3, but 7 cannot be written as a product of factors; therefore, 7 is called a prime number or prime. A prime is a positive whole number that cannot be written as the product of two smaller factors. Numbers 5 and 3 are primes but 4 and 10 are not since $4 = 2 \cdot 2$, and $10 = 2 \cdot 5$. Numbers that can be factored, like 4, 6, and 10, are called composite. The number 1 is not composite but, because it behaves so differently from other numbers, it is not usually considered a prime either; consequently 2 is the first prime, and the first few primes are:

$$2, 3, 5, 7, 11, 13, 17, 19, 23, 29, 31, 37, 41, 43, 47$$

| | | | | | | | | | | | | | | | | | | | |
|---|---|---|---|---|---|---|---|---|---|---|---|---|---|---|---|---|---|---|---|
| 2 | 3 | 5 | 7 | 11 | 13 | 17 | 19 | 23 | 29 | 31 | 37 | 41 | 43 | 47 | 53 | 59 | 61 | 67 | 71 |
| 73 | 79 | 83 | 89 | 97 | 101 | 103 | 107 | 109 | 113 | 127 | 131 | 137 | 139 | 149 | 151 | 157 | 163 | 167 | 173 |
| 179 | 181 | 191 | 193 | 197 | 199 | 211 | 223 | 227 | 229 | 233 | 239 | 241 | 251 | 257 | 263 | 269 | 271 | 277 | 281 |
| 283 | 293 | 307 | 311 | 313 | 317 | 331 | 337 | 347 | 349 | 353 | 359 | 367 | 373 | 379 | 383 | 389 | 397 | 401 | 409 |
| 419 | 421 | 431 | 433 | 439 | 443 | 449 | 457 | 461 | 463 | 467 | 479 | 487 | 491 | 499 | 503 | 509 | 521 | 523 | 541 |
| 547 | 557 | 563 | 569 | 571 | 577 | 587 | 593 | 599 | 601 | 607 | 613 | 617 | 619 | 631 | 641 | 643 | 647 | 653 | 659 |
| 661 | 673 | 677 | 683 | 691 | 701 | 709 | 719 | 727 | 733 | 739 | 743 | 751 | 757 | 761 | 769 | 773 | 787 | 797 | 809 |
| 811 | 821 | 823 | 827 | 829 | 839 | 853 | 857 | 859 | 863 | 877 | 881 | 883 | 887 | 907 | 911 | 919 | 929 | 937 | 941 |
| 947 | 953 | 967 | 971 | 977 | 983 | 991 | 997 | | | | | | | | | | | | |

| | | | | | | | | | | | | | | | | | | | |
|---|---|---|---|---|---|---|---|---|---|---|---|---|---|---|---|---|---|---|---|
| 1009 | 1013 | 1019 | 1021 | 1031 | 1033 | 1039 | 1049 | 1051 | 1061 | 1063 | 1069 | 1087 | 1091 | 1093 | 1097 | 1103 | 1109 | 1117 | 1123 |
| 1129 | 1151 | 1153 | 1163 | 1171 | 1181 | 1187 | 1193 | 1201 | 1213 | 1217 | 1223 | 1229 | 1231 | 1237 | 1249 | 1259 | 1277 | 1279 | 1283 |
| 1289 | 1291 | 1297 | 1301 | 1303 | 1307 | 1319 | 1321 | 1327 | 1361 | 1367 | 1373 | 1381 | 1399 | 1409 | 1423 | 1427 | 1429 | 1433 | 1439 |
| 1447 | 1451 | 1453 | 1459 | 1471 | 1481 | 1483 | 1487 | 1489 | 1493 | 1499 | 1511 | 1523 | 1531 | 1543 | 1549 | 1553 | 1559 | 1567 | 1571 |
| 1579 | 1583 | 1597 | 1601 | 1607 | 1609 | 1613 | 1619 | 1621 | 1627 | 1637 | 1657 | 1663 | 1667 | 1669 | 1693 | 1697 | 1699 | 1709 | 1721 |
| 1723 | 1733 | 1741 | 1747 | 1753 | 1759 | 1777 | 1783 | 1787 | 1789 | 1801 | 1811 | 1823 | 1831 | 1847 | 1861 | 1867 | 1871 | 1873 | 1877 |
| 1879 | 1889 | 1901 | 1907 | 1913 | 1931 | 1933 | 1949 | 1951 | 1973 | 1979 | 1987 | 1993 | 1997 | 1999 | | | | | |

| | | | | | | | | | | | | | | | | | | | |
|---|---|---|---|---|---|---|---|---|---|---|---|---|---|---|---|---|---|---|---|
| 2003 | 2011 | 2017 | 2027 | 2029 | 2039 | 2053 | 2063 | 2069 | 2081 | 2083 | 2087 | 2089 | 2099 | 2111 | 2113 | 2129 | 2131 | 2137 | 2141 |
| 2143 | 2153 | 2161 | 2179 | 2203 | 2207 | 2209 | 2213 | 2221 | 2237 | 2239 | 2243 | 2251 | 2267 | 2269 | 2273 | 2281 | 2287 | 2293 | 2297 |
| 2309 | 2311 | 2333 | 2339 | 2341 | 2347 | 2351 | 2357 | 2371 | 2377 | 2381 | 2383 | 2389 | 2393 | 2399 | 2411 | 2417 | 2423 | 2437 | 2441 |
| 2447 | 2459 | 2467 | 2473 | 2477 | 2503 | 2521 | 2531 | 2539 | 2543 | 2549 | 2551 | 2557 | 2579 | 2591 | 2593 | 2609 | 2617 | 2621 | 2633 |
| 2647 | 2657 | 2659 | 2663 | 2671 | 2677 | 2683 | 2687 | 2689 | 2693 | 2699 | 2707 | 2711 | 2713 | 2719 | 2729 | 2731 | 2741 | 2749 | 2753 |
| 2767 | 2777 | 2789 | 2791 | 2797 | 2801 | 2803 | 2819 | 2833 | 2837 | 2843 | 2851 | 2857 | 2861 | 2879 | 2887 | 2897 | 2903 | 2909 | 2917 |
| 2927 | 2939 | 2953 | 2957 | 2963 | 2969 | 2971 | 2999 | | | | | | | | | | | | |

| | | | | | | | | | | | | | | | | | | | |
|---|---|---|---|---|---|---|---|---|---|---|---|---|---|---|---|---|---|---|---|
| 3001 | 3011 | 3019 | 3023 | 3037 | 3041 | 3049 | 3061 | 3067 | 3079 | 3083 | 3089 | 3109 | 3119 | 3121 | 3137 | 3163 | 3167 | 3169 | 3181 |
| 3187 | 3191 | 3203 | 3209 | 3217 | 3221 | 3229 | 3251 | 3253 | 3257 | 3259 | 3271 | 3299 | 3301 | 3307 | 3313 | 3319 | 3323 | 3329 | 3331 |
| 3343 | 3347 | 3359 | 3361 | 3371 | 3373 | 3389 | 3391 | 3407 | 3413 | 3433 | 3449 | 3457 | 3461 | 3463 | 3467 | 3469 | 3491 | 3499 | 3511 |
| 3517 | 3527 | 3529 | 3533 | 3539 | 3541 | 3547 | 3557 | 3559 | 3571 | 3581 | 3583 | 3593 | 3607 | 3613 | 3617 | 3623 | 3631 | 3637 | 3643 |
| 3659 | 3671 | 3673 | 3677 | 3691 | 3697 | 3701 | 3709 | 3719 | 3727 | 3733 | 3739 | 3761 | 3767 | 3769 | 3779 | 3793 | 3797 | 3803 | 3821 |
| 3823 | 3833 | 3847 | 3851 | 3853 | 3863 | 3877 | 3881 | 3889 | 3907 | 3911 | 3917 | 3919 | 3923 | 3929 | 3931 | 3943 | 3947 | 3967 | 3989 |

| | | | | | | | | | | | | | | | | | | | |
|---|---|---|---|---|---|---|---|---|---|---|---|---|---|---|---|---|---|---|---|
| 4001 | 4003 | 4007 | 4013 | 4019 | 4021 | 4027 | 4049 | 4051 | 4057 | 4073 | 4079 | 4091 | 4093 | 4099 | 4111 | 4127 | 4129 | 4133 | 4139 |
| 4153 | 4157 | 4159 | 4177 | 4201 | 4211 | 4217 | 4219 | 4229 | 4231 | 4241 | 4243 | 4253 | 4259 | 4261 | 4271 | 4273 | 4283 | 4289 | 4297 |
| 4327 | 4337 | 4339 | 4349 | 4357 | 4363 | 4373 | 4391 | 4397 | 4409 | 4421 | 4423 | 4441 | 4447 | 4451 | 4457 | 4463 | 4481 | 4483 | 4493 |
| 4507 | 4513 | 4517 | 4519 | 4523 | 4547 | 4549 | 4561 | 4567 | 4583 | 4591 | 4597 | 4603 | 4621 | 4637 | 4639 | 4643 | 4649 | 4651 | 4657 |
| 4663 | 4673 | 4679 | 4691 | 4703 | 4721 | 4723 | 4729 | 4733 | 4751 | 4759 | 4783 | 4787 | 4789 | 4793 | 4799 | 4801 | 4813 | 4817 | 4831 |
| 4861 | 4871 | 4877 | 4889 | 4903 | 4909 | 4919 | 4931 | 4933 | 4937 | 4943 | 4951 | 4957 | 4967 | 4969 | 4973 | 4987 | 4993 | 4999 | |

| | | | | | | | | | | | | | | | | | | | |
|---|---|---|---|---|---|---|---|---|---|---|---|---|---|---|---|---|---|---|---|
| 5003 | 5009 | 5011 | 5021 | 5023 | 5039 | 5051 | 5059 | 5077 | 5081 | 5087 | 5099 | 5101 | 5107 | 5113 | 5119 | 5147 | 5153 | 5167 | 5171 |
| 5179 | 5189 | 5197 | 5209 | 5227 | 5231 | 5233 | 5237 | 5261 | 5273 | 5279 | 5281 | 5297 | 5303 | 5309 | 5323 | 5333 | 5347 | 5351 | 5381 |
| 5387 | 5393 | 5399 | 5407 | 5413 | 5417 | 5419 | 5431 | 5437 | 5441 | 5443 | 5449 | 5471 | 5477 | 5479 | 5483 | 5501 | 5503 | 5507 | 5519 |
| 5521 | 5527 | 5531 | 5557 | 5563 | 5569 | 5573 | 5581 | 5591 | 5623 | 5639 | 5641 | 5647 | 5651 | 5653 | 5657 | 5659 | 5669 | 5683 | 5689 |
| 5693 | 5701 | 5711 | 5717 | 5737 | 5741 | 5743 | 5749 | 5779 | 5783 | 5791 | 5801 | 5807 | 5813 | 5821 | 5827 | 5839 | 5843 | 5849 | 5851 |
| 5857 | 5861 | 5867 | 5869 | 5879 | 5881 | 5897 | 5903 | 5923 | 5927 | 5939 | 5953 | 5981 | 5987 | | | | | | |

| | | | | | | | | | | | | | | | | | | | |
|---|---|---|---|---|---|---|---|---|---|---|---|---|---|---|---|---|---|---|---|
| 6007 | 6011 | 6029 | 6037 | 6043 | 6047 | 6053 | 6067 | 6073 | 6079 | 6089 | 6091 | 6101 | 6113 | 6121 | 6131 | 6133 | 6143 | 6151 | 6163 |
| 6173 | 6197 | 6199 | 6203 | 6211 | 6217 | 6221 | 6229 | 6247 | 6257 | 6263 | 6269 | 6271 | 6277 | 6287 | 6299 | 6301 | 6311 | 6317 | 6323 |
| 6329 | 6337 | 6343 | 6353 | 6359 | 6361 | 6367 | 6373 | 6379 | 6389 | 6397 | 6421 | 6427 | 6449 | 6451 | 6469 | 6473 | 6481 | 6491 | 6521 |
| 6529 | 6547 | 6551 | 6553 | 6563 | 6569 | 6571 | 6577 | 6581 | 6599 | 6607 | 6619 | 6637 | 6653 | 6659 | 6661 | 6673 | 6679 | 6689 | 6691 |
| 6701 | 6703 | 6709 | 6719 | 6733 | 6737 | 6761 | 6763 | 6779 | 6781 | 6791 | 6793 | 6803 | 6823 | 6827 | 6829 | 6833 | 6841 | 6857 | 6863 |
| 6869 | 6871 | 6883 | 6899 | 6907 | 6911 | 6947 | 6949 | 6959 | 6961 | 6967 | 6971 | 6977 | 6983 | 6991 | 6997 | | | | |

| | | | | | | | | | | | | | | | | | | | |
|---|---|---|---|---|---|---|---|---|---|---|---|---|---|---|---|---|---|---|---|
| 7001 | 7013 | 7019 | 7027 | 7039 | 7043 | 7057 | 7069 | 7079 | 7103 | 7109 | 7121 | 7127 | 7129 | 7151 | 7159 | 7177 | 7187 | 7193 | 7207 |
| 7211 | 7213 | 7219 | 7229 | 7237 | 7243 | 7247 | 7253 | 7283 | 7297 | 7307 | 7309 | 7321 | 7331 | 7333 | 7349 | 7351 | 7369 | 7393 | 7411 |
| 7417 | 7433 | 7451 | 7457 | 7459 | 7477 | 7481 | 7487 | 7489 | 7499 | 7507 | 7517 | 7523 | 7529 | 7537 | 7541 | 7547 | 7549 | 7559 | 7561 |
| 7573 | 7577 | 7583 | 7589 | 7591 | 7603 | 7607 | 7621 | 7639 | 7643 | 7649 | 7669 | 7673 | 7681 | 7687 | 7691 | 7699 | 7703 | 7717 | 7723 |
| 7727 | 7741 | 7753 | 7757 | 7759 | 7789 | 7793 | 7817 | 7823 | 7829 | 7841 | 7853 | 7867 | 7873 | 7877 | 7879 | 7883 | 7901 | 7907 | 7919 |
| 7927 | 7933 | 7937 | 7949 | 7951 | 7963 | 7993 | | | | | | | | | | | | | |

**Figure 3.1**  A chart showing the first 1008 prime numbers. A color chart showing the first 1291 primes is available from Creative Publications, Inc. P.O. Box 10328, Palo Alto, CA 94303.

A glance reveals that this sequence does not follow any simple law. In fact, the structure of the sequence of primes is extremely complicated.

The prime numbers are scarce when we consider large number ranges. For example, there are

168 prime numbers between 1 and 1000
135 prime numbers between 1000 and 2000

128 prime numbers between 2000 and 3000
120 prime numbers between 3000 and 4000
119 prime numbers between 4000 and 5000

The reason for this is clear: the larger a number is, the more numerous its potential divisors, and the less likely it is to be a prime. Nevertheless, the list of prime numbers appears to be endless. In fact, Euclid proved that there is an infinite number of primes.

Let us now look at our first important problem in number theory: How can one decide whether a number is a prime?

One method of finding a prime is to divide the given number by all numbers less than the number. For example, assume you want to determine if 751 is prime. You can successively divide by 2, 3, 4, 5, 6, 7, . . . . If no exact division occurs, you know that 751 is a prime number. If an exact division does occur, you have found a factor of 751.

One of the first questions to arise concerns the number of divisions necessary. Is it necessary to divide by every number from 1 through 751? The answer, fortunately, is no! You need only to determine whether any number less than or equal to $\sqrt{751}$ divides 751.

**Example 1.**   Is 91 a prime? $\sqrt{91} = 9 +$; by trying the numbers 1, 2, 3, 4, 5, . . ., one sees that $91 = 7 \cdot 13$.

**Example 2.**   Is 1973 a prime? $\sqrt{1973} = 44+$. Since no number less than or equal to 43 divides 1973, this number is prime.

In general terms, if the factors of N are 1, 2, 3, 4, . . ., N, the whole range is uncovered by dividing by:

$$1, 2, 3, 4, \ldots, \sqrt{N}$$

For large numbers this method may be very cumbersome; however, here as in many other calculations of number theory, you can rely on modern computational techniques. It is simple to program a computer to divide a given number by all integers up to the square root of the number and to print those which given no remainder.

The following BASIC program uses this technique to produce all primes less than 400. Since we know immediately and automatically that 2 is the only even prime, the program examines only the odd numbers starting with

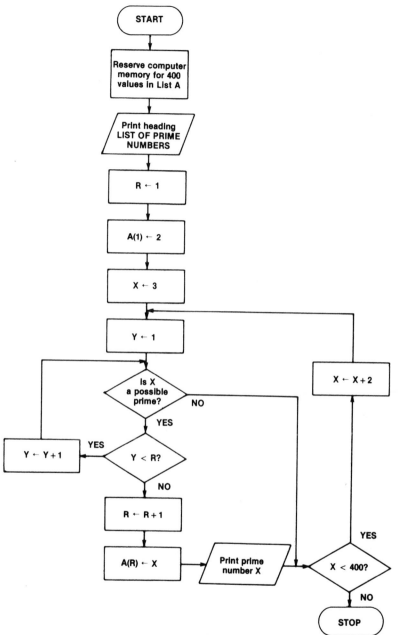

**Figure 3.2**  Prime number flowchart.

the number 3, and divides each succeeding odd number up to $\sqrt{400}$ by all primes that are found. A flowchart is shown in Figure 3.2.

```
10  REM PRIME NUMBER GENERATOR
12  DIM A[400]
15  PRINT "PRIME NUMBERS"
20  LET R=1
25  LET A[1]=2
30  LET P=1
35  FOR X=3 TO 400 STEP 2
40  FOR Y=1 TO R
45  IF INT(X/A[Y])*A[Y]=X THEN 95
50  NEXT Y
55  LET R=R+1
60  LET A[R]=X
65  IF P>6 THEN 85
70  LET P=P+1
75  PRINT X;
80  GOTO 95
85  LET P=1
90  PRINT X
95  NEXT X
99  END

RUN

PRIME NUMBERS
  3    5    7   11   13   17   19
 23   29   31   37   41   43   47
 53   59   61   67   71   73   79
 83   89   97  101  103  107  109
113  127  131  137  139  149  151
157  163  167  173  179  181  191
193  197  199  211  223  227  229
233  239  241  251  257  263  269
271  277  281  283  293  307  311
313  317  331  337  347  349  353
359  367  373  379  383  389  397
```

The first statement in the program is a REM statement giving the name of the program. The second statement causes the computer to reserve 400 locations for list A. Statement 30 will cause the heading LIST OF PRIME NUMBERS to be printed in columns 1 through 21 on the teletypewriter. Statement 40 sets variable $R$ to 1 and the first value in the list, A(1), to 2. The group of statements beginning with the FOR statement at line number 60 and ending with the statement at line number 130 is repeated for $X = 3$ to $X = 399$, $X$ increasing by 2 with each iteration. This loop contains another loop starting with the statement at line number 70 and ending with the statement at line 90. In this small loop, the variable $X$ is tested by the INT function to determine whether it is divisible by previously found prime numbers. Whenever $X$ is not divisible by a previous prime, a remainder is lost when performing the operation INT($X$/A($Y$)). When this result is multiplied by A($Y$), a number less than the original value of $X$ is obtained. The IF-THEN statement compares the number (INT($X$/A($Y$)) * A($Y$)) to $X$ and proceeds to the statement at line number 90 if the number is less than $X$.

Iteration continues in the smaller loop until $Y = R$, or until a division occurs without remainder. When division without remainder occurs, $X$ is not a possible prime, so the program transfers control directly to the statement at line number 130, and iteration of the larger loop begins again with $X$ increased by 2.

If no branch to the statement at line number 130 occurs, $X$ is a prime number. In this case $R$ is increased by 1, the value of $X$ is stored in A($R$), and the prime number is printed. When the value of $X$ equals 399, the computation is completed and the program terminates.

Now look at a BASIC program that will determine whether a given positive integer is a prime number. In this program, one types the number on a terminal. The program determines whether $N$ is prime and so indicates in an output message. Program and sample results are shown below.

```
100  REM IS THE NUMBER PRIME ?
110  PRINT "WHAT IS THE NUMBER";
120  INPUT N
130  IF INT(N)=N THEN 160
140  PRINT N;"IS NOT AN INTEGER"
150  GOTO 250
160  IF N >= 2 THEN 190
170  PRINT N;"IS LESS THAN 2"
180  GOTO 250
190  FOR I=2 TO SQR(N)
```

```
200  IF INT(N/I)=N/I THEN 240
210  NEXT I
220  PRINT N;"IS A PRIME NUMBER"
230  GOTO 250
240  PRINT N;"IS NOT A PRIME NUMBER"
250  PRINT
260  PRINT "TYPE 1 TO CONTINUE; 2 TO STOP";
270  INPUT C
280  IF C=1 THEN 110
290  END

RUN

WHAT IS THE NUMBER?624
 624 IS NOT A PRIME NUMBER

TYPE 1 TO CONTINUE; 2 TO STOP?1
WHAT IS THE NUMBER?769
 769  IS A PRIME NUMBER

TYPE 1 TO CONTINUE; 2 TO STOP?1
WHAT IS THE NUMBER?1
 1    IS LESS THAN 2

TYPE 1 TO CONTINUE; 2 TO STOP?1
WHAT IS THE NUMBER?76.34
 76.34     IS NOT AN INTEGER

TYPE 1 TO CONTINUE; 2 TO STOP?1
WHAT IS THE NUMBER?953
 953  IS A PRIME NUMBER

TYPE 1 TO CONTINUE; 2 TO STOP?2
```

To determine if the input number is prime this program attempts to divide that number by all possible integers between 1 and the number. The program checks the input number to see if it is an integer greater than 1. Appropriate messages are printed if the input number is not an integer, or is a number less than 2.

## 3.2   THE SIEVE OF ERATOSTHENES

There exists an ancient method of finding prime numbers, known as the sieve of Eratosthenes. Eratosthenes (276–194 B.C.E.) was a Greek scholar, chief librarian of the famous library in Alexandria. He is noted for his chronology of ancient history and for his measurement of the meridian between Assuan and Alexandria, which made it possible to estimate the dimensions of the earth with fairly great accuracy.

# ERATOSTHENES

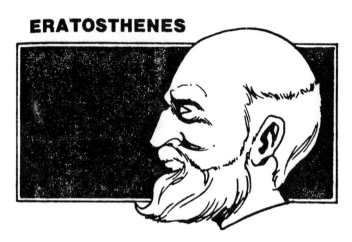

Eratosthenes's sieve method consists of writing down all numbers up to some limit, say 50:

| | | | | | | |
|---|---|---|---|---|---|---|
| 2 | 3 | 4 | 5 | ~~6~~ | 7 | ~~8~~ |
| ~~9~~ | ~~10~~ | 11 | ~~12~~ | 13 | ~~14~~ | ~~15~~ |
| ~~16~~ | 17 | ~~18~~ | 19 | ~~20~~ | ~~21~~ | ~~22~~ |
| 23 | ~~24~~ | ~~25~~ | ~~26~~ | ~~27~~ | ~~28~~ | 29 |
| ~~30~~ | 31 | ~~32~~ | ~~33~~ | ~~34~~ | ~~35~~ | ~~36~~ |
| 37 | ~~38~~ | ~~39~~ | ~~40~~ | 41 | ~~42~~ | 43 |
| ~~44~~ | ~~45~~ | ~~46~~ | 47 | ~~48~~ | ~~49~~ | ~~50~~ |

We begin with the number 2, and cross out every number that is divisible by 2, except 2 itself. Next, we cross out all the multiples of 3, except 3 itself. 4 can be ignored, since it is already crossed out. 5 is next; we cross out all multiples of 5 except 5 itself. We next cross out all multiples of 7, and all remaining numbers must be primes. It is not necessary to consider numbers greater than $\sqrt{50}$, such as 11, 13, and so forth, because any multiple of these numbers that is in our list, for example 22, must have a factor smaller than 7, and therefore must have already been crossed out.

The sieve of Eratosthenes can be used to construct prime tables. At the Los Alamos Scientific Laboratory all primes up to 100 000 000 have been stored in computer memory. Tables have been printed listing primes up to 10 000 000. Although prime tables beyond 10 000 000 can easily be produced by using computers, there does not seem to be much point in going to the considerable expense and effort to have them printed. Rarely does a mathematician, even a specialist in number theory, need to decide whether a very large number is a prime.

The following BASIC program computes all the prime numbers up to 1000 using the sieve of Eratosthenes.

```
100   REM SIEVE OF ERATOSTHENES
110   DIM N[1000],P[200]
120   FOR I=2 TO 1000
130   LET N[I]=0
140   NEXT I
150   LET K=0
160   FOR P=2 TO 1000
170   IF N[P]<0 THEN 240
180   LET K=K+1
190   LET P[K]=P
200   IF P>SQR(1000) THEN 240
210   FOR I=P TO 1000 STEP P
220   LET N[I]=-1
230   NEXT I
240   NEXT P
250   REM PRINT PRIME NUMBERS
260   LET C=1
270   FOR I=1 TO K
280   PRINT P[I];
290   LET C=C+1
300   IF C<= 7 THEN 330
310   PRINT
320   LET C=1
330   NEXT I
340   END

RUN
```

| 2   | 3   | 5   | 7   | 11  | 13  | 17  |
|-----|-----|-----|-----|-----|-----|-----|
| 19  | 23  | 29  | 31  | 37  | 41  | 43  |
| 47  | 53  | 59  | 61  | 67  | 71  | 73  |
| 79  | 83  | 89  | 97  | 101 | 103 | 107 |
| 109 | 113 | 127 | 131 | 137 | 139 | 149 |
| 151 | 157 | 163 | 167 | 173 | 179 | 181 |
| 191 | 193 | 197 | 199 | 211 | 223 | 227 |
| 229 | 233 | 239 | 241 | 251 | 257 | 263 |
| 269 | 271 | 277 | 281 | 283 | 293 | 307 |
| 311 | 313 | 317 | 331 | 337 | 347 | 349 |
| 353 | 359 | 367 | 373 | 379 | 383 | 389 |
| 397 | 401 | 409 | 419 | 421 | 431 | 433 |
| 439 | 443 | 449 | 457 | 461 | 463 | 467 |
| 479 | 487 | 491 | 499 | 503 | 509 | 521 |
| 523 | 541 | 547 | 557 | 563 | 569 | 571 |
| 577 | 587 | 593 | 599 | 601 | 607 | 613 |
| 617 | 619 | 631 | 641 | 643 | 647 | 653 |
| 659 | 661 | 673 | 677 | 683 | 691 | 701 |
| 709 | 719 | 727 | 733 | 739 | 743 | 751 |
| 757 | 761 | 769 | 773 | 787 | 797 | 809 |
| 811 | 821 | 823 | 827 | 829 | 839 | 853 |
| 857 | 859 | 863 | 877 | 881 | 883 | 887 |
| 907 | 911 | 919 | 929 | 937 | 941 | 947 |
| 953 | 967 | 971 | 977 | 983 | 991 | 997 |

The program uses two arrays. Array N is to be thought of as the list of numbers from 2 to 1000, with N($I$) standing for $I$. The program crosses out the number $I$ by setting N($I$) $= -1$. Array P contains the primes. The number $K$ indicates how many primes we have found. The program is initialized by zeroing out array N and setting $K$ to 0. Then the program performs the sieve (statements 160 through 240). The program picks the next number $P$ in array N. If it has been crossed out, it is forgotten. If not, it is added to array P and all of its multiples are crossed out. Finally the program prints the list of prime numbers.

## 3.3 MERSENNE PRIMES

The Mersenne primes are prime numbers of the special form

$$M_p = 2^p - 1,$$

where $p$ is another prime. These numbers came into mathematics early and appear in Euclid's discussion of the perfect numbers, which we shall encounter in Chapter 4. They are named for the French friar Marin Mersenne (1588-1648), who performed calculations upon and with perfect numbers.

When you start calculating the numbers for various primes $p$ you immediately see that they are not all primes. For example,

$$M_2 = 2^2 - 1 = 3 = \text{prime},$$

$$M_3 = 2^3 - 1 = 7 = \text{prime},$$

$$M_5 = 2^5 - 1 = 31 = \text{prime},$$

$$M_7 = 2^7 - 1 = 127 = \text{prime},$$

$$M_{11} = 2^{11} - 1 = 2047 = \text{not prime} (23 \times 89)$$

The general procedure for finding large primes of the Mersenne type is to examine all the numbers $M_p$ for the various primes $p$. The numbers increase very rapidly and so do the labors involved. The reason why the work is manageable even for quite large numbers is that there are very effective ways for determining whether these special numbers are primes.

There was an early phase in the examination of Mersenne primes which culminated in 1750 when the Swiss mathematician Buler established that $M_{31}$ is a prime. By that time only eight Mersenne primes had been found. In 1876 the French mathematician Lucas established that the 39-digit number

$$M_{127} = 170\ 141\ 183\ 460\ 469\ 231\ 731\ 687\ 303\ 715\ 884\ 105\ 727$$

is a prime. The introduction of electronic hand calculators made it possible to find Mersenne primes up to $M_{257}$, but the results were disappointing; no further Mersenne primes were found.

This was the situation when computers came into being. With the development of newer and larger computers it was possible to push the search for Mersenne primes further and further. By 1963, Mersenne primes had been computed up to $M_{11213}$, a prime number containing 3376 digits. This prime was computed with a computer at the University of Illinois. In 1971, the primacy of $M_{19737}$ was determined, and in 1978, the primacy of $M_{21701}$ was determined. In February 1979, $M_{23209}$ was determined to be the twenty-sixth Mersenne prime.

The following BASIC program computes several Mersenne primes by using a simple plug-in technique. The program data supplies the prime numbers. The recursion relationship is guaranteed to produce primes when prime numbers are used as exponents.

```
100   REM MERSENNE PRIMES
110   PRINT "PRIME", "MERSENNE"
120   PRINT "NUMBER", "PRIMES"
130   PRINT
140   FOR K=1 TO 8
150   READ P
160   LET M=2↑P-1
170   PRINT P,M
180   NEXT K
190   DATA 2,3,5,7,13,17,19
200   END

RUN

PRIME               MERSENNE
NUMBER              PRIMES
  2                   3
  3                   7
  5                  31
  7                 127
 13                8191
 17              131071
 19              524287
```

As discussed in Chapter 1, the world's largest known prime number was discovered by David Slowinski and Harry Nelson, computer scientists at the Lawrence Livermore National Laboratory in California. This 13395 digit number is far too large to have any meaning in the physical world. For example, it would take the fastest computer in the world more than 100 000 years even to count to a 21-digit number. In April 1979, the twenty-seventh Mersenne prime, $M_{44497}$, was found during the acceptance testing of a CRAY-1 supercomputer at the Lawrence Livermore National Laboratory (see Figure 3.3). Historically, the largest known prime number has been a Mersenne prime. Mathematicians and computer scientists will continue their efforts to find larger and larger prime numbers, but until the next Mersenne prime number is found, $M_{44497}$ will remain the largest known prime number in the world.

# THE 27$^{TH}$ MERSENNE PRIME

The 27th Mersenne prime. It has 13395 digits and equals $2^{44497} - 1$.

**Figure 3.3** The 27th Mersenne prime and the largest known prime number in the world, $M_{44497}$. This 13 395 digit number was discovered in 1979 by two scientists at Lawrence Livermore National Laboratory using a CRAY-1 supercomputer. A color chart of this number is available from Creative Publications, Inc. P.O. Box 10328, Palo Alto, CA 94303.

## 3.4 TWIN PRIMES

With the exception of 2 the primes are odd, so any two consecutive primes must have a distance that is greater than or equal to 2. Pairs of primes with the shortest distance (2) are called twin primes. For example, the following number pairs are twin primes:

3, 5     5, 7     11, 13     17, 19     29, 31     10 006 427, 10 006 429

It has been proven that there are infinitely many prime numbers, however, we still do not know whether there are an infinite number of twin primes. Consequently, mathematicians have been looking for large twin primes for centuries.

The following BASIC program computes and prints all twin primes less than 1000.

```
100    REM TWIN PRIMES LESS THAN 1000
110    DIM A[1000],B[400]
120    FOR X=2 TO 1000
130    LET A[X]=0
140    NEXT X
150    LET C=0
160    LET S=SQR(1000)
170    FOR B=2 TO 1000
180    IF A[B]<0 THEN 250
190    LET C=C+1
200    LET B[C]=B
210    IF B>S THEN 250
220    FOR X=B TO 1000 STEP B
230    LET A[X]=-1
240    NEXT X
250    NEXT B
260    PRINT "TWIN PRIMES"
270    PRINT
280    FOR X=2 TO C
290    IF B[X]-B[X-1] <> 2 THEN 310
300    PRINT B[X-1];B[X]
310    NEXT X
320    END

RUN
```

TWIN PRIMES

| | |
|---|---|
| 3 | 5 |
| 5 | 7 |
| 11 | 13 |
| 17 | 19 |
| 29 | 31 |
| 41 | 43 |
| 59 | 61 |
| 71 | 73 |
| 101 | 103 |
| 107 | 109 |
| 137 | 139 |
| 149 | 151 |
| 179 | 181 |
| 191 | 193 |
| 197 | 199 |
| 227 | 229 |
| 239 | 241 |
| 269 | 271 |
| 281 | 283 |
| 311 | 313 |
| 347 | 349 |
| 419 | 421 |
| 431 | 433 |
| 461 | 463 |
| 521 | 523 |
| 569 | 571 |
| 599 | 601 |
| 617 | 619 |
| 641 | 643 |
| 659 | 661 |
| 809 | 811 |
| 821 | 823 |
| 827 | 829 |
| 857 | 859 |

## 3.5 PRIME POLYNOMINALS

No formula exists that produces all possible primes, but several remarkable expressions produce large numbers of primes for consecutive values of $x$. For example, $2x^2 + 29$ will give primes (starting with 29) for $x = 0$ to 28 (twenty-nine primes), $x^2 + 41$ will give primes for $x = 0$ to 39 (forty primes starting with 41); $x^2 + x + 17$ will generate sixteen primes, $6x^2 + 6x + 31$ will give primes for twenty-nine values of $x$, $3x^2 + 3x + 23$ will give primes for twenty-two values of $x$, and $x^2 - 79x + 1601$ will give eighty consecutive prime values when $x = 0, 1, 2, \ldots, 79$. Other examples of the same nature exist.

The following BASIC program uses the formula

$$x^2 - x + 41$$

to generate a prime for the 41 values of $x$: 0, 1, 2, 3, $\ldots$, 40.

```
100  REM PRIME POLYNOMIAL
110  PRINT "PRIME NUMBERS"
120  FOR X=1 TO 40
130  LET P=X↑2-X+41
140  PRINT P,
150  NEXT X
160  END

RUN
```

PRIME NUMBERS

| | | | | |
|------|------|------|------|------|
| 41   | 43   | 47   | 53   | 61   |
| 71   | 83   | 97   | 113  | 131  |
| 151  | 173  | 197  | 223  | 251  |
| 281  | 313  | 347  | 383  | 421  |
| 461  | 503  | 547  | 593  | 641  |
| 691  | 743  | 797  | 853  | 911  |
| 971  | 1033 | 1097 | 1163 | 1231 |
| 1301 | 1373 | 1447 | 1523 | 1601 |

## Review Exercises

1. Which of the following numbers are primes?

   (a) The year of your birth?
   (b) The present year number?
   (c) Your house number?
   (d) Your girlfriend's/boyfriend's age?

2. Find the prime factors of 333, of 459, of 589, of 703.

3. Tabulate the number of primes for each hundred: 1–100, 101–200, up to 901–1000.

4. Identify a pair of prime numbers that differ by 1. Why are they the only such pair?

5. Use a sieve of Eratosthenes to hand calculate the prime numbers less than 200.

6. Modify the sieve of Eratosthenes program so that it will produce primes less than 2000.

7. Write a BASIC program that will generate prime numbers from the polynomial $x^2 + x + 41$ when $x$ varies from 0 to 40.

8. The polynomial $x^2 + x + 17$ may be used to generate sixteen prime numbers when $x = 0, 1, 2, \ldots, 15$. Write a BASIC program that will produce a printout of these sixteen prime numbers.

9. Write a program to determine if 15 485 863 is a prime number.

10. The year 1951 will be remembered as a mathematical oddity because 1951 is a prime number. Draw a flowchart and write a program to find all the prime numbers between 1951 and 2000.

11. Modify the BASIC program shown on page 146 to compute the first 1000 prime numbers.

12. Two consecutive odd prime numbers, such as 5 and 7, or 17 and 19, are called twin primes. The existence or nonexistence of an infinite number of such number pairs is still one of the unsolved problems of number theory. The number of prime triplets, such as 3, 5, 7, however, has been proven to be finite. Write a BASIC program that will produce several prime triplets.

# Chapter 4

# MAGIC CURIOS

**Preview**

A magic number curio is a problem like the following: find digits $a$, $b$, $c$, $d$ such that

$$a^b \times c^d = abcd$$

An answer to this problem is

$$2^5 \times 9^2 = 2592$$

Some number curios appear to be occurring purely by chance. Many others are due to peculiarities in the decimal number system and do not occur when the same numbers are transformed to a base other than 10. Still others have much significance in the field of number theory.

After you complete this chapter, you should be able to:

1. Identify perfect numbers, amicable numbers, Armstrong numbers, lucky numbers, abundant numbers, deficient numbers, and square numbers.
2. See how a computer can be used to produce perfect numbers, amicable numbers, Armstrong numbers, and square numbers.
3. See how a computer can determine if a given number is abundant, deficient, or perfect.

## 4.1 PERFECT NUMBERS

The number 6 has a curious property. By adding the divisors of 6, a sum equal to the number itself is found:

$$3 + 2 + 1 = 6$$

This is also true of the number 28.

$$14 + 7 + 4 + 2 + 1 = 28$$

These numbers are called perfect. A formula for generating perfect numbers is

$$2^{n-1}(2^n - 1)$$

in which the factor $2^n - 1$ must be a Mersenne prime. Today, several perfect numbers are known, those computed by setting $n$ equal to 2, 3, 5, 7, 13, 17, 19, 31, 61, 89, 107, 127, 521, 607, 1279, 2203, 2281, 3217, 4253, 4423, 9689, 9941, 11213, or 19937 in the previous formula. Expanded, the first eight are 6, 28, 496, 8128, 33 550 336, 8 589 869 056, 137 438 691 328, and 2 305 843 008 139 952 128.

No one knows whether an odd number can be perfect. None has been found, yet no one has proved they do not exist. In 1968, Bryant Tuckerman at IBM announced that an odd perfect number must have a least 36 digits. Several interesting facts exist about perfect numbers. Every even perfect number must end in either 28 or 6. The sum of the reciprocals of all the divisors of an even perfect number must equal 2. For example, the perfect number 6 has divisors 1, 2, 3, and 6, and their reciprocals total $1/1 + 1/2 + 1/3 + 1/6 = 2$.

In recent years, computers have been used to generate very large perfect numbers. In the 1950s, a computer was used to determine the thirteenth through seventeenth perfect numbers: $2^{520}(2^{521} - 1)$ through $2^{2280}(2^{2281} - 1)$. The latter number contains 1372 digits. As discussed in Chapter 1, an 18-year-old high school student used computers to compute the twenty-first through twenty-third perfect numbers. The latter number contains 7723 digits. The twenty-fourth perfect number when expanded has 12 003 digits.

The following BASIC program computes the first two perfect numbers,

```
10   FOR N=2 TO 100
20   LET S=0
30   FOR D=1 TO N/2
40   IF INT(N/D) <> N/D THEN 60
50   LET S=S+D
60   NEXT D
70   IF S <> N THEN 90
```

```
80  PRINT N;"IS A PERFECT NUMBER"
90  NEXT N
99  END


RUN


 6   IS A PERFECT NUMBER
28   IS A PERFECT NUMBER
```

The previous program was executed on a TRS-80 microcomputer. Greater accuracy can be obtained by executing the program on larger machines. The following results are correct to nineteen decimal places. They were obtained by executing a program on a large Control Data computer.

| $n$ | $2^n - 1$ | $2^{n-1}$ | Perfect Number |
|---|---|---|---|
| 13 | 8191 | 4096 | 33550336 |
| 17 | 131071 | 65536 | 8589869056 |
| 19 | 524287 | 262144 | 137438691328 |
| 31 | 2147483647 | 1073741824 | 2305843008139952128 |

## 4.2  AMICABLE NUMBERS

Once there was a king who thought of himself as quite a mathematician. He told a prisoner, "Give me a problem to solve. You may go free until I solve it. But as soon as I have the answer, off comes your head!" Now the prisoner was rather clever himself. Here is the problem he gave the king. 220 and 284 are called amicable numbers. The sum of the proper divisors of 220 equals 284,

$$1 + 2 + 4 + 5 + 10 + 11 + 20 + 22 + 44 + 55 + 110 = 284$$

and the sum of the proper divisors of 284 equals 220.

$$1 + 2 + 4 + 71 + 142 = 220$$

Find the next pair of amicable numbers! The story goes that the prisoner went free and finally died of old age. The king never solved the problem.

There are about 400 amicable pairs of numbers known, of which some are:

| | | |
|---:|:---:|:---|
| 220 | and | 284 (the smallest pair) |
| 1184 | | 1210 |
| 2620 | | 2924 |
| 5020 | | 5564 |
| 6232 | | 6368 |
| 10744 | | 10856 |
| 12285 | | 14595 |
| 17296 | | 18416 |
| 63020 | | 76084 |
| 66928 | | 66992 |
| 67095 | | 71145 |
| 69615 | | 87633 |
| 79750 | | 88730 |
| 9 363 584 | | 9 437 056 |
| 111 448 537 712 | | 118 853 793 424 |

Several methods are available for finding amicable pairs. One common method is to let

$$A = (3)(2^x) - 1$$
$$B = (3)(2^{x-1}) - 1$$
$$C = (9)(2^{2x-1}) - 1$$

If $x$ is greater than 1, and A, B, and C are all primes, then $2^x$AB and $2^x$C constitute an amicable pair of numbers. For example, if $x = 4$, then A = 47, B = 23, and C = 1151, which are all primes. Then

$$(2^4)(47)(23) = 17\ 296$$

and

$$(2^4)(1151) = 18\ 416$$

Too bad the king didn't have a computer. He could have written the following BASIC program to produce the next pair of amicable numbers.

```
100 REM AMICABLE NUMBERS
110 FOR A = 1 TO 7000
120 LET S = 0
130 FOR D = 1 TO A/2
```

```
140 IF A/D <>INT(A/D) THEN 160
150 LET S=S+D
160 NEXT D
170 IF S <= A THEN 260
180 LET B = S
190 LET T = 0
200 FOR F=1 TO B/2
210 IF B/F <>INT(B/F) THEN 230
220 LET T=T+F
230 NEXT F
240 IF T<>A THEN 260
250 PRINT A;"AND";B;"ARE AMICABLE NUMBERS"
260 NEXT A
270 END

RUN

  220  AND 284  ARE AMICABLE NUMBERS
  1184    AND 1210    ARE AMICABLE NUMBERS
```

There are a total of five pairs of amicable numbers less than 10 000. A scheme for finding them is as follows:

• Start with a number, $N$
• Factor it, and find the sum, $S$, of its factors, counting 1 as a factor.
• If $S$ is more than $N$, find $S1$, the sum of the factors of $S$.
• If $S1 = N$, print the amicable pair: $N$, $S$.
• Only even values of $N$ are used, since all amicable numbers are even.

Let's use this scheme on the pair: 220, 284.

• $N = 220$
• Sum of factors of 220 is 284. Thus, $S = 284$.
• Since $S(284)$ is more than $N(220)$, find $S1$ (sum of factors of 284). $S1 = 220$.
• $S1(220) = N(220)$. Therefore, 220 and 284 are amicable numbers.

The reader should draw a flowchart and write a BASIC program using this procedure.

The search for amicable pairs is eminently suited to computers. For each number $N$ let the computer determine all divisors ($\neq N$) and their sum $M$. Then perform the same operation on $M$. If you return to the original number $N$ by this procedure, an amicable pair $(N, M)$ has been discovered. Several years ago, this procedure was used on a computer at Yale with $N$s of up to one million resulted in the collection of 42 pairs of amicable numbers.

Actually we know very little about the properties of the amicable numbers, but on the basis of the numbers shown one can make some conjectures. For instance, it appears that the quotient of the two numbers must get closer and closer to 1 as they increase. From the list of numbers given you can see that both numbers may be even or both odd, but no case has been found in which one was odd and the other even.

### 4.3 ARMSTRONG NUMBERS

One hundred fifty three is an interesting number because

$$153 = 1^3 + 5^3 + 3^3$$

Numbers such as this are called Armstrong numbers. Any $N$ digit number is an Armstrong number if the sum of the $N^{th}$ power of the digits is equal to the original number.

The following program finds three other three-digit Armstrong numbers.

```
100   REM ARMSTRONG NUMBERS
110   FOR N=100 TO 999
120   LET A=INT(N/100)
130   LET B=INT(N/10)-10*A
140   LET C=N-100*A-10*B
150   IF N <> A↑3+B↑3+C↑3 THEN 190
160   PRINT "ARMSTRONG NUMBER";N
170   PRINT "EQUALS";A↑3;"+";B↑3;"+";C↑3
180   PRINT
190   NEXT N
200   END

RUN

ARMSTRONG NUMBER 153
EQUALS  1    + 125  + 27
```

```
ARMSTRONG NUMBER 370
EQUALS  27   + 343  + 0

ARMSTRONG NUMBER 371
EQUALS  27   + 343  +1

ARMSTRONG NUMBER 407
EQUALS  64   + 0    + 343
```

## 4.4  SPECIAL FOUR-DIGIT NUMBERS

The four-digit number 3025 is special. The sum of the first two digits (30) and the last two digits (25) is 55. If you square 55 the original number is obtained ($55^2 = 3025$).

The following BASIC program determines all four-digit numbers that have this property.

```
10 REM SPECIAL FOUR-DIGIT NUMBERS
20 PRINT "N","SUM","SUM SQUARED"
30 PRINT
40 FOR N = 1000 TO 9999
50 LET F=INT(N/100)
60 LET L=N-100*F
70 IF(F+L)↑2 <>N THEN 90
80 PRINT N, F+L,(F+L)↑2
90 NEXT N
99 END

RUN
```

| N | SUM | SUM SQUARED |
|---|-----|-------------|
| 2025 | 45 | 2025 |
| 3025 | 55 | 3025 |
| 9801 | 99 | 9801 |

## 4.5  LUCKY NUMBERS

A group of investigators working with Stanislav M. Ulam at Los Alamos Scientific Laboratories have discovered what they call the lucky numbers, determined by a sieving process. As with the sieve of Eratosthenes, we begin by writing down all the natural numbers, in order, limiting ourselves to the first hundred to illustrate the process. If we leave 1 and strike out every second number, we eliminate all the even numbers.

| 1 | 3 | 5̸ | 7 | 9 |
|---|---|---|---|---|
| 1̸1̸ | 13 | 15 | 1̸7̸ | 1̸9̸ |
| 21 | 2̸3̸ | 25 | 2̸7̸ | 2̸9̸ |
| 31 | 33 | 3̸5̸ | 37 | 3̸9̸ |
| 4̸1̸ | 43 | 4̸5̸ | 4̸7̸ | 49 |
| 51 | 5̸3̸ | 5̸5̸ | 5̸7̸ | 5̸9̸ |
| 6̸1̸ | 63 | 6̸5̸ | 67 | 69 |
| 7̸1̸ | 73 | 75 | 7̸7̸ | 79 |
| 8̸1̸ | 8̸3̸ | 8̸5̸ | 87 | 8̸9̸ |
| 9̸1̸ | 93 | 9̸5̸ | 9̸7̸ | 99 |

In Eratosthenes' sieve we next struck out every multiple of 3 because 3 was the next surviving number. The rule here is different: strike out every third number among those remaining. That means that 5 goes, and 11, 17, 23, etc. All such numbers are crossed out by a single slant line. The next surviving number is 7, so we let that stand and cross out every seventh remaining one (19, 39, etc.) with two slant lines, to indicate what is happening. Then cross out every 9th, then every 13th, and so on. The slant lines indicate at what stage in the construction each number was eliminated.

An excellent exercise for the reader is to use this procedure to produce all lucky numbers less than 1000.

## 4.6  ABUNDANT AND DEFICIENT NUMBERS

The number the sum of whose divisors is less than the number itself is called deficient, and a number exceeded by this sum is called abundant. As discussed in Section 4.1, the number is perfect when the sum of the divisors of that number, excluding the number itself, equals the number in question.

For example,

$$6 = 1 + 2 + 3 \text{ and is perfect.}$$

$$12 < 1 + 2 + 3 + 4 + 6 \text{ and is abundant.}$$

$$10 > 1 + 2 + 5 \text{ and is deficient.}$$

The following BASIC program factors a given number into its divisors and determines whether the number is abundant, deficient, or perfect.

```
100   REM ABUNDANT AND DEFICIENT NUMBERS
110   PRINT "THIS PROGRAM WILL TAKE A NUMBER AND"
120   PRINT "COMPUTE THE SUM OF ITS DIVISORS"
130   PRINT
140   PRINT "TYPE THE NUMBER";
150   INPUT N
160   LET S=0
170   PRINT "THE DIVISORS OF";N;"ARE";
180   FOR X=1 TO N-1
190   IF N/X <> INT(N/X) THEN 220
200   LET S=S+X
210   PRINT X;
220   NEXT X
230   PRINT
240   IF S>N THEN 280
250   IF S<N THEN 300
260   PRINT N;"IS PERFECT"
270   GOTO 310
280   PRINT N;"IS ABUNDANT"
290   GOTO 310
300   PRINT N;"IS DEFICIENT"
310   PRINT
320   PRINT "TYPE 1 TO CONTINUE; 2 TO STOP";
330   INPUT Z
340   IF Z=1 THEN 130
350   END

RUN

THIS PROGRAM WILL TAKE A NUMBER AND
COMPUTE THE SUM OF ITS DIVISORS

TYPE THE NUMBER?12
THE DIVISORS OF 12    ARE 1     2      3      4      6
   12   IS ABUNDANT

TYPE 1 TO CONTINUE; 2 TO STOP?1
```

```
TYPE THE NUMBER?6
THE DIVISORS OF 6    ARE 1    2    3
  6    IS PERFECT

TYPE 1 TO CONTINUE; 2 TO STOP?1

TYPE THE NUMBER?15
THE DIVISORS OF 15   ARE 1    3    5
 15    IS DEFICIENT

TYPE 1 TO CONTINUE; 2 TO STOP?2
```

## 4.7  SQUARE NUMBERS

Numbers like 4, 9, 16, 25, 36, etc. are called square numbers because

$$2^2 = 4$$
$$3^2 = 9$$
$$4^2 = 16$$
$$5^2 = 25$$
$$6^2 = 36$$

and so on.

Certain pairs of numbers when added or subtracted give a square number. For example: 8 and 17

$$8 + 17 = 25 \text{ (a square number)}$$
$$17 - 8 = \phantom{0}9 \text{ (a square number)}$$

The following program finds all the pairs of numbers less than 100 that give a square number when added and when subtracted.

```
100   REM SQUARE NUMBERS
110   PRINT " N      P    N+P    P-N"
120   PRINT
130   FOR N=1 TO 100
140   FOR P=N+1 TO 100
150   IF SQR(N+P) <> INT(SQR(N+P)) THEN 180
```

```
160  IF SQR(P-N) <> INT(SQR(P-N)) THEN 180
170  PRINT N;P;N+P;P-N
180  NEXT P
190  NEXT N
200  END
```

RUN

| N | P | N+P | P-N |
|---|---|-----|-----|
| 4 | 5 | 9 | 1 |
| 6 | 10 | 16 | 4 |
| 8 | 17 | 25 | 9 |
| 10 | 26 | 36 | 16 |
| 12 | 13 | 25 | 1 |
| 12 | 37 | 49 | 25 |
| 14 | 50 | 64 | 36 |
| 16 | 20 | 36 | 4 |
| 16 | 65 | 81 | 49 |
| 18 | 82 | 100 | 64 |
| 20 | 29 | 49 | 9 |
| 24 | 25 | 49 | 1 |
| 24 | 40 | 64 | 16 |
| 28 | 53 | 81 | 25 |
| 30 | 34 | 64 | 4 |
| 32 | 68 | 100 | 36 |
| 36 | 45 | 81 | 9 |
| 36 | 85 | 121 | 49 |
| 40 | 41 | 81 | 1 |
| 42 | 58 | 100 | 16 |
| 48 | 52 | 100 | 4 |
| 48 | 73 | 121 | 25 |
| 54 | 90 | 144 | 36 |
| 56 | 65 | 121 | 9 |
| 60 | 61 | 121 | 1 |
| 64 | 80 | 144 | 16 |
| 70 | 74 | 144 | 4 |
| 72 | 97 | 169 | 25 |
| 80 | 89 | 169 | 9 |
| 84 | 85 | 169 | 1 |
| 96 | 100 | 196 | 4 |

## Review Exercises

1. Show that 496 is a perfect number.

2. Show that 762 is not a perfect number.

3. Show that 8128 is a perfect number.

4. Prepare a talk or paper on how a computer can be used to generate perfect numbers.

5. Explain why no prime number can be perfect.

6. Draw a flowchart that could have been used to write the perfect number program shown in Section 4.1.

7. The number 25 can be written as the sum of two squares: $3^2 + 4^2 = 25$. Write a BASIC program to find all numbers less than 50 which can be written as the sum of two squares.

8. If the sum of $1 + 2 + 3 \ldots + k$ is a perfect square ($N^2$) and if $N$ is less than 100, what are the possible values of $k$? For example,

    If $k = 8$
    $1 + 2 + 3 + 4 + 5 + 6 + 7 + 8 = 36$
    36 is a perfect square; $36 = 6^2$ and $6 < 100$.

    Write a BASIC program to find other possible values for $k$.

9. Draw a flowchart that could have been used to write the amicable numbers program shown in Section 4.2.

10. A certain number is divisible by 13. When this number is divided by the numbers 2 through 12, there is always a remainder of 1. Write a BASIC program to determine the smallest number that fits these conditions.

11. You were walking down Broadway in New York City and found a piece of paper. Written on the paper was a set of clues for a missing number:

    When this number is divided by 10, there is a remainder of 9.
    When divided by 9 there is a remainder of 8.
    When divided by 7 there is a remainder of 6.
    When divided by 6 there is a remainder of 5.
    When divided by 5 there is a remainder of 4.
    When divided by 4 there is a remainder of 3.

When divided by 3 there is a remainder of 2.
When divided by 2 there is a remainder of 1.

Write a BASIC program that will determine the smallest number that fits these clues.

12. Draw a flowchart and write a BASIC program to generate all lucky numbers less than 1000.

# Chapter 5

# FACTORING

**Preview**

In Chapter 3 we discussed prime numbers with little reference to factors. The two are closely related since most methods of finding primes also infer factorization. For example, the number 12 is not prime because the factors of 12 are

1, 2, 3, 4, 6, and 12.

Factors, then, are those integers that can divide into a number exactly.
   After completing this chapter, you should be able to:

1.   Define the factors, the greatest common factor, and the least common multiple of numbers.
2.   See how a computer can be used to compute the factors of a number, the greatest common divisor of two numbers, and the least common multiple of three numbers.
3.   Calculate the GCD and LCM of sets of numbers.
4.   Write a BASIC program to find the factors of a number.

## 5.1   FACTORIZATION

In Chapter 3 we saw that every integer greater than 1 is either a prime number or a composite number. In this section we shall find that every integer greater than 1 can be expressed in terms of its prime factors essentially in only one way.

Consider the various ways of factoring 24:

$$24 = 1 \times 24;$$
$$24 = 2 \times 12;$$
$$24 = 3 \times 8;$$
$$24 = 4 \times 6;$$
$$24 = 2 \times 2 \times 6;$$
$$24 = 2 \times 3 \times 4;$$
$$24 = 2 \times 2 \times 2 \times 3 = 2^3 \times 3.$$

The last factorization in terms of the prime numbers 2 and 3 could be written as $2 \times 3 \times 2^2$ and in other ways; however, these ways are equivalent since the order of the factors does not affect the product. Thus 24 can be expressed in terms of its prime factors in one and only one way.

One of the easiest ways to find the prime factors of a number is to consider the prime numbers

$$2, 3, 5, 7, 11, 13, 17, 19, 23, 29, 31, 37, \ldots$$

in order and use each as a factor as many times as possible. For 24 we would have

$$24 = 2 \times 12$$
$$= 2 \times 2 \times 6$$
$$= 2 \times 2 \times 2 \times 3$$

You can also write these steps using division.

$$2 \overline{)24}$$
$$2 \overline{)12}$$
$$2 \underline{)\ 6}$$
$$3$$

Since 3 is a prime number, no further steps are needed; $24 = 2^3 \times 3$.

**Example.**    Express 4680 in terms of its prime factors.

$$
\begin{array}{r}
2)\overline{4680} \\
2)\overline{2340} \\
2)\overline{1170} \\
3)\ \overline{585} \\
3)\ \overline{195} \\
5)\ \ \overline{65} \\
\overline{13}
\end{array}
\qquad 4680 = 2^3 \times 3^2 \times 5 \times 13
$$

The following BASIC program will find the prime factors of any integer. The integer is input to the program as data. The program terminates execution whenever a zero is typed as input.

```
100  REM PRIME FACTORS OF ANY INTEGER
110  PRINT "PRIME FACTORS OF ANY INTEGER"
120  PRINT
130  PRINT
140  PRINT
150  PRINT "NUMBER TO BE FACTORED IS";
160  INPUT A
170  IF ABS(A) <= 1 THEN 340
180  LET N= INT(ABS(A))
190  REM FIND AND PRINT PRIMES
200  LET B=0
210  FOR I=2 TO N/2
220  IF N/I>INT(N/I) THEN 300
230  LET B=B+1
240  IF B>1 THEN 260
250  PRINT "PRIME FACTORS OF"; N; "ARE"
260  PRINT I;
270  LET N=N/I
280  IF N=1 THEN 120
290  LET I=I-1
300  NEXT I
310  IF N <> INT(A) THEN 120
320  PRINT N;"IS A PRIME NUMBER"
330  GOTO 130
340  END
```

```
RUN

PRIME FACTORS OF ANY INTEGER


NUMBER TO BE FACTORED IS?56
PRIME FACTORS OF 56   ARE
  2    2    2    7


NUMBER TO BE FACTORED IS?346
PRIME FACTORS OF 346 ARE
  2    173


NUMBER TO BE FACTORED IS?397
  397  IS A PRIME NUMBER


NUMBER TO BE FACTORED IS? 560
PRIME FACTORS OF 560   ARE
  2    2    2    2    5    7
```

In the sample program RUN, five integers are factored, including one prime number and one integer with two fairly large prime factors.

In 1643, Pierre de Fermat (1601-1665), a French mathematical genius, illustrated a method of factoring numbers that did not require divisions. His method was based on the following procedure:

Assume    $N = A * B$ where $A \leq B$.

Assume    $N$, $A$, and $B$ are all odd integers.

Let    $X = (A + B)/2$ and,

   $Y = (B - A)/2$.

Then    $N = A^2 - B^2$.

The method involves searching for values of $A$ and $B$ that satisfy these equations where $0 < B < A$ and $A < B \leq N$.

The following BASIC program uses Fermat's method to compute the largest factor of a given integer.

```
100  REM LARGEST FACTOR OF ANY NUMBER
110  PRINT "WHAT IS THE NUMBER";
120  INPUT N
130  IF N=0 THEN 280
140  LET W=INT(SQR(N))
150  LET X=2*W+1
160  LET Y=1
170  LET R=W*W-N
180  IF R=0 THEN 250
190  IF R>0 THEN 220
200  LET R=R+X
210  LET X=X+2
220  LET R=R-Y
230  LET Y=Y+2
240  GOTO 180
250  LET F=(X-Y)/2
260  PRINT "LARGEST FACTOR OF";N;"IS";F
270  GOTO 110
280  END

RUN

WHAT IS THE NUMBER?311
LARGEST FACTOR OF 311   IS 1
WHAT IS THE NUMBER?45
LARGEST FACTOR OF 45    IS 5
WHAT IS THE NUMBER?0
```

The BASIC program shown below computes the largest factor of the numbers listed in a DATA statement.

```
100  REM LARGEST FACTOR PROGRAM
110  READ N
120  FOR D=2 TO SQR(N)
130  IF N/D=INT(N/D) THEN 170
140  NEXT D
```

```
150   PRINT N,"IS A PRIME NUMBER"
160   GOTO 110
170   PRINT N/D,"IS THE LARGEST FACTOR OF ";N
180   GOTO 110
190   DATA 3394,5799,2827,1907,9115
200   DATA 2807,1495,373,19,206
210   END

RUN
```

| | | |
|---|---|---|
| 1697 | IS THE LARGEST FACTOR OF | 3394 |
| 1933 | IS THE LARGEST FACTOR OF | 5799 |
| 257 | IS THE LARGEST FACTOR OF | 2827 |
| 1907 | IS A PRIME NUMBER | |
| 1823 | IS THE LARGEST FACTOR OF | 9115 |
| 401 | IS THE LARGEST FACTOR OF | 2807 |
| 299 | IS THE LARGEST FACTOR OF | 1495 |
| 373 | IS A PRIME NUMBER | |
| 19 | IS A PRIME NUMBER | |
| 103 | IS THE LARGEST FACTOR OF | 206 |

```
OUT OF DATA  IN LINE 110
```

To find all pairs of factors of an integer, use the following BASIC program.

```
100   REM PAIRS OF FACTORS OF AN INTEGER
110   PRINT "PAIRS OF FACTORS"
120   PRINT
130   PRINT
140   PRINT "TYPE THE INTEGER";
150   INPUT X
160   PRINT
170   PRINT "THE PAIRS OF FACTORS OF";X;"ARE:"
180   FOR A=1 TO SQR(ABS(X))
190   IF INT(X/A) <> X/A THEN 210
200   PRINT A,X/A
210   NEXT A
220   PRINT
```

```
230  PRINT
240  PRINT "TYPE 1 TO STOP; 2 TO CONTINUE";
250  INPUT T
260  IF T <> 1 THEN 120
270  END

RUN

PAIRS OF FACTORS


TYPE THE INTEGER?8960

THE PAIRS OF FACTORS OF 8960    ARE:
   1            8960
   2            4480
   4            2240
   5            1792
   7            1280
   8            1120
  10             896
  14             640
  16             560
  20             448
  28             320
  32             280
  35             256
  40             224
  56             160
  64             140
  70             128
  80             112


TYPE 1 TO STOP; 2 TO CONTINUE?2


TYPE THE INTEGER?4680

THE PAIRS OF FACTORS OF 4680 ARE:
   1            4680
   2            2340
```

| | |
|---|---|
| 3 | 1560 |
| 4 | 1170 |
| 5 | 936 |
| 6 | 780 |
| 8 | 585 |
| 9 | 520 |
| 10 | 468 |
| 12 | 390 |
| 13 | 360 |
| 15 | 312 |
| 18 | 260 |
| 20 | 234 |
| 24 | 195 |
| 26 | 180 |
| 30 | 156 |
| 36 | 130 |
| 39 | 120 |
| 40 | 117 |
| 45 | 104 |
| 52 | 90 |
| 60 | 78 |
| 65 | 72 |

```
TYPE 1 TO STOP; 2 TO CONTINUE?1
```

## 5.2  GREATEST COMMON DIVISOR

The problem of factoring is closely related to finding the greatest common divisor (GCD) of two numbers.

If $A$ and $B$ are two integers, any number that divides both $A$ and $B$ is called a common divisor of A and B. For example, the numbers that divide 12 are 1, 2, 3, 4, 6, and 12. The numbers that divide 18 are 1, 2, 3, 6, 9, 18. Then the common divisors of 12 and 18 are 1, 2, 3, 6. The largest of the divisors is called the greatest common divisor (GCD) of $A$ and $B$. In the previous example, 6 is the greatest common divisor (GCD) of 12 and 18.

We may use the prime factorization of two numbers to find their greatest common divisor. Express each number by its prime factorization, consider the prime numbers that are factors of both of the given numbers, and take the product of those prime numbers with each raised to the highest power that is a factor of both of the given numbers. For $12 = 2^2 \times 3$ and $18 = 2 \times 3^2$ we have GCD $= 2 \times 3$; that is, 6.

**Example.** Find the greatest common divisor of 3850 and 5280.

$$3850 = 2 \times 5^2 \times 7 \times 11;$$
$$5280 = 2^5 \times 3 \times 5 \times 11.$$

The GCD of 3850 and 5280 is $2 \times 5 \times 11$; that is, 110.

Even though the GCD of two numbers can be found quite easily by using the previous method, there exists an ancient method that is more suited for computer computation. A procedure known as Euclid's algorithm (which occurs in the seventh book of Euclid's Elements—about 300 B.C.E.) is one of the basic methods of elementary number theory. Suppose that $A$ and $B$ are the two numbers. We divide $B$ into $A$, using integer division, finding a quotient $Q$ and a remainder $R$. This means that

$$A = Q * B + R.$$

Then we divide $R$ into $B$ and keep iterating until the remainder is 0. The last nonzero divisor becomes the greatest common divisor.

**Example 1.** Let us find the GCD of 1976 and 1032.

$$1976 = 1032 \times 1 + 944$$
$$1032 = 944 \times 1 + 88$$
$$944 = 88 \times 10 + 56$$
$$88 = 56 \times 1 + 32$$
$$56 = 32 \times 1 + 24$$
$$32 = 24 \times 1 + 8$$
$$24 = 8 \times 3 + 0$$

The GCD of 1976 and 1032 is 8.

**Example 2.** Find the GCD of 76084 and 63020.

$$76084 = 63020 \times 1 + 13064$$
$$63020 = 13064 \times 4 + 10764$$

$$13064 = 10764 \times 1 + 2300$$
$$10764 = 2300 \times 4 + 1564$$
$$2300 = 1564 \times 1 + 736$$
$$1564 = 736 \times 2 + 92$$
$$736 = 92 \times 8 + 0$$

The GCD of 76084 and 63020 is 92.

The following BASIC program uses Euclid's algorithm to compute the GCD of a given pair of numbers.

```
100   REM FIND THE GCD OF TWO NUMBERS
120   PRINT"    A","   B"," GCD"
130   PRINT
140   READ A,B
150   PRINT A,B,
160   LET Q=INT(A/B)
170   LET R=A-Q*B
180   LET A=B
190   LET B=R
200   IF R>0 THEN 160
210   PRINT A
220   GOTO 140
230   DATA 60,5280,49,139,3850,5280
240   DATA 1124,1472,17296,18416
250   DATA 76084.,63020.,7854,13398
260   END

RUN
```

| A | B | GCD |
|---|---|---|
| 60 | 5280 | 60 |
| 49 | 139 | 1 |
| 3850 | 5280 | 110 |
| 1124 | 1472 | 4 |
| 17296 | 18416 | 16 |
| 76084. | 63020. | 92 |
| 7854 | 13398 | 462 |

```
OUT OF DATA  IN LINE 140
```

In this sample RUN, we use the algorithm to compute the GCD of seven pairs of numbers.

## 5.3  LEAST COMMON MULTIPLE

The set of multiples of 12 is 12, 24, 36, 48, 60, 72, 84, 96, 108, ...; the set of multiples of 18 is 18, 36, 54, 72, 90, 108, ...; the set of common multiples of 12 and 18 is 36, 72, 108, .... Notice that the least of these common multiples, 36, is a divisor of each of the common multiples. In general, the common multiple of two numbers which is a divisor of each of the common multiples is called the least common multiple (LCM) of the two numbers.

We may use the prime factorization of two numbers to find their least common multiple. Express each number by its prime factorization, consider the prime factors that are factors of either of the given numbers, and take the product of these prime numbers with each raised to the highest power that occurs in either of the prime factorizations. For $12 = 2^2 \times 3$ and $18 = 2 \times 3^2$, we have LCM $= 2^2 \times 3^2$; that is, 36.

**Example.**  Find the least common multiple of 3850 and 5280.

$$3850 = 2 \times 5^2 \times 7 \times 11;$$
$$5280 = 2^5 \times 3 \times 5 \times 11.$$

The least common multiple of 3850 and 5280 is $2^5 \times 3 \times 5^2 \times 7 \times 11$; that is 184 800.

A common multiple of three numbers $A$, $B$, and $C$ is a number divisible by all of them. Among these multiples there is a least common multiple. The LCM of 2, 3, and 6 is 6. The LCM of 5, 10, and 20 is 20. The LCM of 12, 18, 24 is 72. A general technique for finding the LCM of three numbers follows:

1. Identify the numbers $A$, $B$, $C$.
2. Arbitrarily let $X = A$.
3. Does $A$ divide $X$?
4. If yes, go to Step 6.
5. If no, increase $X$ by 1 and return to Step 3.
6. Does $B$ divide $X$?
7. If yes, go to Step 9.
8. If no, increase $X$ by 1 and return to Step 3.
9. Does $C$ divide $X$?

10. If yes, go to Step 12.
11. If no, increase $X$ by 1 and return to Step 3.
12. Print the value of $X$.

This procedure is illustrated in flowchart form in Figure 5.1. A BASIC program to compute the LCM of three numbers follows.

In this sample RUN, the program determines the least common multiple of six sets of numbers.

```
100   REM LEAST COMMON MULTIPLE
110   READ A,B,C
120   LET X=A
130   IF INT(X/A)=X/A THEN 160
140   LET X=X+1
150   GOTO 130
160   IF INT(X/B)=X/B THEN 190
170   LET X=X+1
180   GOTO 130
190   IF INT(X/C)=X/C THEN 220
200   LET X=X+1
210   GOTO 130
220   PRINT "THE LCM OF ";A;B;C;"IS";X
230   GOTO 110
240   DATA 5,7,23,15,25,54,24,10,17
250   DATA 24,63,99,16,24,62,5,10,15
260   END

RUN

THE LCM OF 5    7    23   IS 805
THE LCM OF 15   25   54   IS 1350
THE LCM OF 24   10   17   IS 2040
THE LCM OF 24   63   99   IS 5544
THE LCM OF 16   24   62   IS 1488
THE LCM OF 5    10   15   IS 30

OUT OF DATA  IN LINE 110
```

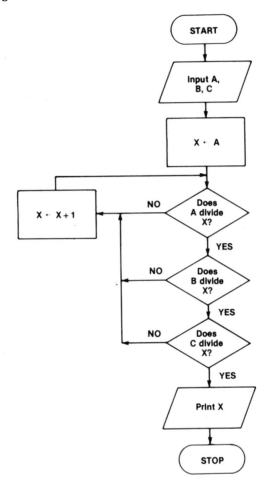

**Figure 5.1**   LCM of three numbers.

## 5.4   *GCD* AND *LCM* FOR SEVERAL NUMBERS

So far the greatest common divisor and the least common multiple have been explained only for two and three numbers, respectively, but there is no difficulty in extending these concepts. Let us consider first the case of three numbers *A*, *B*, and *C*. A common divisor is any number dividing them all. Among these common divisors there is a greatest common divisor. For example, let us determine the GCD of the three numbers 76 084, 63 020, and 196.

In a previous example we have already found that the GCD of 76 084 and 63 020 is 92; consequently the GCD of 92 and 196 is 4. Therefore, the GCD of 76 084, 63 020, and 92 is 4.

To find the LCM of the three numbers 24, 18, and 52, you calculate the LCM of 24 and 18, which is 72, and then the LCM of 72 and 52, which results in 936. Therefore, the LCM of 24, 18, and 52 is 936.

All the properties associated with producing the GCD and LCM for three numbers can be used to produce the GCD and LCM for any set of numbers.

The following BASIC program finds the GCD and LCM of a set of numbers.

```
100   REM LCM AND GCD OF FOUR INTEGERS
110   PRINT "THIS PROGRAM COMPUTES THE LCM AND"
120   PRINT "GCD OF FOUR INTEGERS"
130   PRINT
140   PRINT
150   PRINT "TYPE THE NUMBERS";
160   INPUT A[1],A[2],A[3],A[4]
170   IF A[1]=0 THEN 530
180   LET K=A[1]
190   REM FIND LARGEST NUMBER
200   FOR I=1 TO 4
210   IF K >= A[1] THEN 230
220   LET K=A[1]
230   NEXT I
240   LET L=1
250   LET G=1
260   REM COMPUTE THE LCM AND GCD
270   FOR J=2 TO K
280   FOR I=1 TO 4
290   LET B[I]=1
300   NEXT I
310   REM DETERMINE HOW MANY J'S ARE IN EACH INT
320   FOR I=1 TO 4
330   IF A[I]/J <> INT(A[I]/J) THEN 370
340   LET A[I]=A[I]/J
350   LET B[I]=B[I]*J
370   NEXT I
380   REM MAX NO. OF J'S IN NUMBERS
381   REM MIN NO. OF J'S COMMON TO ALL NUMBERS
```

```
390   LET F=B[1]
400   LET M=B[1]
410   FOR I=2 TO 4
420   IF F >= B[I] THEN 440
430   LET F=B[I]
440   IF M <= B[I] THEN 460
450   LET M=B[I]
460   NEXT I
470   LET L=L*F
480   LET G=G*M
490   NEXT J
500   PRINT "THE LEAST COMMON MULTIPLE IS";L
510   PRINT "THE GREATEST COMMON DIVISOR IS";G
520   GOTO 130
530   END

RUN

THIS PROGRAM COMPUTES THE LCM AND
GCD OF FOUR INTEGERS

TYPE THE NUMBERS?16,24,62,120
THE LEAST COMMON MULTIPLE IS 240
THE GREATEST COMMON DIVISOR IS 2


TYPE THE NUMBERS?28,12,64,88
THE LEAST COMMON MULTIPLE IS 14784
THE GREATEST COMMON DIVISOR IS 4


TYPE THE NUMBERS?36,24,49,104
THE LEAST COMMON MULTIPLE IS 45864.
THE GREATEST COMMON DIVISOR IS 1
```

The program printout shows the LCM and GCD for five sets of four numbers.

**Review Exercises**

1.  Find the prime factorization of each integer:

    (a)   68       (d)   738
    (b)   76       (e)   819
    (c)  123       (f)  1425

2.  Hand calculate the GCD of 68 and 76.

3.  Hand calculate the GCD of 60 and 5280.

4.  Using Euclid's algorithm, determine the GCD of 49 and 139.

5.  Solve Exercise 2 by means of Euclid's algorithm.

6.  Find the GCD of each of the first four pairs of amicable numbers. Check the results against those obtained from the prime factorizations.

7.  Write a BASIC program to determine the greatest common divisor of 100 and 350, 720 and 820, 17 and 25.

8.  Draw a flowchart that could have been used to write the greatest common divisor program shown on page 143.

9.  Hand calculate the LCM of 68 and 76.

10.  Hand calculate the LCM of 76 and 1425.

11.  Hand calculate the LCM of 5, 23, and 91.

12.  Find the LCM for each of the first four pairs of amicable numbers.

13.  The number 10 has four factors: 1, 2, 5, and 10. They all divide evenly into 10. The number 48 has ten factors: 1, 2, 3, 4, 6, 8, 12, 16, 24, and 48. They all divide evenly into 48. Write a BASIC program to compute the smallest natural number that has exactly 32 factors.

14.  Write a BASIC program that will divide 1000 into two parts so that one part is a multiple of 19 and the other part is a multiple of 47.

15.  Write a BASIC program to compute the following numbers.

    a.   Smallest number with 1 factor.
    b.   Smallest number with 2 factors.
    c.   Smallest number with 3 factors.

    d.   Smallest number with 4 factors.

    e.   Smallest number with 5 factors.

    f.   Smallest number with 6 factors.

    g.   Smallest number with 7 factors.

    h.   Smallest number with 8 factors.

    i.   Smallest number with 9 factors.

    j.   Smallest number with 10 factors.

16. Write a BASIC program to find the sum of all numbers between 100 and 1000 that are divisible by 14.

17. Write a BASIC program to determine what value of $X$ would make $X434X0$ divisible by 36.

18. Draw a flowchart and write a BASIC program to compute and print the LCM and GCD of several sets of six numbers.

19. Hand calculate the LCM and GCD of the numbers 16, 24, 62, and 120.

# Chapter 6

# FIBONACCI NUMBERS

**Preview**

The only European outstanding for mathematical activity during the Middle Ages was Leonardo Fibonacci of Pisa. This wealthy Italian merchant was fascinated with numbers and discovered, among many other works, a series of numbers known as the Fibonacci series.

After completing this chapter, you should be able to:

1. Identify the Fibonacci number sequence.
2. See how a computer can be used to produce the Fibonacci number sequence.
3. Write a BASIC program to generate Fibonacci numbers.

This must be Fibonacci's apartment!

## 6.1   THE FIBONACCI SEQUENCE

A man bought a pair of rabbits and bred them. The pair produced one pair of young after one month, and a second pair after the second month. Then they stopped breeding. Each new pair also produced two more pairs in the same way, and then stopped breeding. How many new pairs of rabbits did he get each month?

To answer this question, write down in a line the number of pairs in each generation. First write the number 1 for the single pair he started with. Next write the number 1 for the pair they produced after a month.

## FIBONACCI

The next month both pairs had young, so the next number is 2. We now have three numbers in a line: 1, 1, 2, each number representing a new generation. Now the first generation stopped producing. The second generation (1 pair) produced 1 pair, and the third generation (2 pairs) produced 2 pairs, so the next number we write is 1 + 2, or 3. Now the second generation stopped producing. The third generation (2 pairs) produced 2 pairs, and the fourth generation (3 pairs) produced 3 pairs, so the next number we write is 2 + 3 or 5.

Each month, only the last two generations produced, so we can get the next number by adding the last two numbers in the line. The numbers we get this way are called Fibonacci numbers. The first sixteen are 1, 1, 2, 3, 5, 8, 13, 21, 34, 55, 89, 144, 233, 377, 610, and 987. They have interesting properties and appear frequently in nature (see Figure 6.1).

The Fibonacci numbers are such that, after the first two, every number in the sequence equals the sum of the two previous numbers:

$$F_n = F_{n-1} + F_{n-2}$$

Suppose a tree grows according to the following, not unrealistic, formula. Each old branch (including the trunk) puts out one new branch per year; each new branch grows through the next year without branching, after which it qualifies as an old branch. The growth is represented schematically in Figure 6.2. The number of branches after $n$ years is $F_n$.

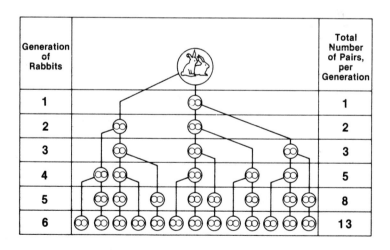

**Figure 6.1**   Rabbits and generations.

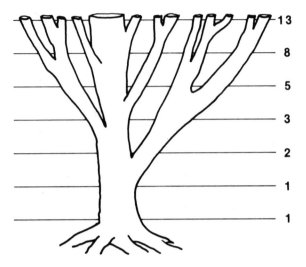

**Figure 6.2**  The Fibonacci tree.

Figure 6.3 illustrates a flowchart of a method that will compute Fibonacci numbers. A BASIC program that computes and prints 30 Fibonacci numbers follows.

```
100   REM FIBONACCI NUMBERS
110   DIM F[30]
120   PRINT "FIBONACCI NUMBERS"
130   PRINT
140   LET F[1]=1
150   LET F[2]=1
160   FOR N=1 TO 28
170   LET F[N+2]=F[N+1]+F[N]
180   NEXT N
190   REM PRINT 30 FIBONACCI NUMBERS
200   FOR X=1 TO 30
210   PRINT F[X]
220   NEXT X
230   END

RUN
```

```
FIBONACCI NUMBERS

        1
        1
        2
        3
        5
        8
       13
       21
       34
       55
       89
      144
      233
      377
      610
      987
     1597
     2584
     4181
     6765
    10946
    17711
    28657
    46368
    75025
   121393
   196418
   317811
   514229
   832040
```

The first statement in the program, a REM statement, gives the name of the program. Statement 110 causes the computer to reserve 30 locations in computer memory for array F. Statement 120 will cause a heading FIBONACCI NUMBERS to be printed. Statements 140 and 150 set variables F(1) and F(2) to 1. The three-statement loop starting at statement 160 causes statement 170 to be executed twenty-eight times. In this loop, twenty-eight Fibonacci numbers are computed. Statement 190 is a REM statement.

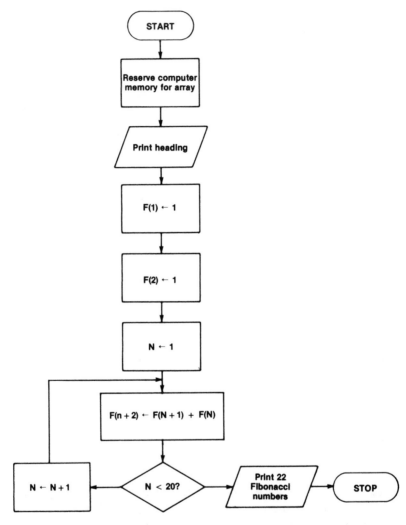

**Figure 6.3**  Fibonacci numbers.

Statements 200, 210, and 220 cause the first thirty Fibonacci numbers to be printed. The END statement terminates the program.

## 6.2    FIBONACCI NUMBERS AND PRIMES

The BASIC program in the last section produced the first thirty Fibonacci numbers. As you have observed, all of these Fibonacci numbers are positive integer quantities, and some are prime numbers.

Let us now consider the problem of generating Fibonacci numbers and identifying those that are primes. An outline of the program procedure follows.

1.  Set F1 and F2 to 1 (F1 is the first Fibonacci number ($F_{n-2}$) and F2 is the second Fibonacci number ($F_{n-1}$)).
2.  Print F1 and F2, identifying each as a prime number.
3.  Perform the following calculations for $I = 3, 4, \ldots, 25$.
    a.  Calculate a value for F using the formula F = F1 + F2.
    b.  Test to see if F is a prime number.
    c.  If F is a prime, identify it as such.
    d.  Update F1 and F2 in preparation for calculating a new Fibonacci number (assign the current value of F1 to F2, then assign the current value of F to F1).

A BASIC program corresponding to the previous procedure follows.

```
100  REM FIBONACCI AND PRIME NUMBERS
110  PRINT "HOW MANY FIBONACCI NUMBERS";
120  INPUT N
130  PRINT
140  PRINT
150  PRINT "FIBONACCI AND PRIME NUMBERS"
160  PRINT
170  LET F1=1
180  LET F2=1
190  PRINT "I=";1,"F=";1;" (PRIME NUMBER)"
200  PRINT "I=";2,"F=";1;" (PRIME NUMBER)"
210  FOR I=3 TO N
220  LET F=F1+F2
230  FOR J=2 TO F-1
240  LET Q=F/J
```

```
250   LET Q1=INT(Q)
260   IF Q=Q1 THEN 300
270   NEXT J
280   PRINT "I=";I,"F=";F;" (PRIME NUMBER)"
290   GOTO 310
300   PRINT "I=";I,"F=";F
310   LET F2=F1
320   LET F1=F
330   NEXT I
340   END
```

RUN

HOW MANY FIBONACCI NUMBERS?24


FIBONACCI AND PRIME NUMBERS

```
I= 1          F= 1          (PRIME NUMBER)
I= 2          F= 1          (PRIME NUMBER)
I= 3          F= 2          (PRIME NUMBER)
I= 4          F= 3          (PRIME NUMBER)
I= 5          F= 5          (PRIME NUMBER)
I= 6          F= 8
I= 7          F= 13         (PRIME NUMBER)
I= 8          F= 21
I= 9          F= 34
I= 10         F= 55
I= 11         F= 89         (PRIME NUMBER)
I= 12         F= 144
I= 13         F= 233        (PRIME NUMBER)
I= 14         F= 377
I= 15         F= 610
I= 16         F= 987
I= 17         F= 1597       (PRIME NUMBER)
I= 18         F= 2584
I= 19         F= 4181
I= 20         F= 6765
I= 21         F= 10946
I= 22         F= 17711
I= 23         F= 28657      (PRIME NUMBER)
I= 24         F= 46368
```

Notice that the program contains a nest of loops. The purpose of the inner loop (statements 130–170) is to determine whether or not each Fibonacci number is prime; the outer loop (statements 110–230) causes the desired sequence of Fibonacci numbers to be computed.

The program generates the first twenty-four Fibonacci numbers, ten of which are primes.

### Review Exercises

1. The following BASIC program will generate the first twenty-two Fibonacci numbers. Execute this program on a computer and compare the printed output with that produced by the program given in Section 6.1.

```
100  REM GENERATE 22 FIBONACCI NUMBERS
105  DIM A[22]
110  LET A[1]=1
120  PRINT A[1]
130  LET A[2]=1
140  PRINT A[2]
150  FOR J=3 TO 22
160  LET A[J]=A[J-1]+A[J-2]
170  PRINT A[J]
180  NEXT J
190  END
```

2. Modify the program in Section 6.1 so that the Fibonacci numbers will be printed more compactly.

3. The Fibonacci numbers are generated by letting the first two numbers of the sequence equal 1; from that point each number may be found by taking the sum of the previous two elements in the sequence. So you get 1, 1, 2, 3, 5, 8, 13, etc. Prepare two lists: one with the first ten and the other with the second ten. For each element from two to nineteen find the difference between the square of the element and the product of the elements immediately preceding and following. In other words, print $F(I)^2 - F(I-1) \times F(I+1)$.

4. Draw a flowchart that could have been used to write the program in Section 6.2.

5. Modify the program in Section 6.2 so that it will generate and print the first thirty-five Fibonacci numbers.

# Chapter 7

# MAGIC SQUARES

## Preview

The popularity of mathematics as a means of recreation and pleasure is evidenced by the frequency with which it is found in popular magazines and newspapers. In this chapter we shall explore one of the most interesting of all mathematical recreations, the magic square.

So much has been written about the subject that it would appear that stagnation is imminent. Still, this entertainment exerts a virtually unbreakable hold on the average game enthusiast or recreational mathematician. Generating magic squares by computer is relatively new and interesting. Not only are magic squares fun to construct, but they provide excellent programming exercises for mathematics students. Once you understand the methods of constructing various magic squares you will find them quite simple to produce, with or without a computer.

After you complete this chapter, you should be able to:

1. Recognize the following types of magic squares: odd order, even order, multiplication, geometric, and prime.
2. Hand calculate magic squares.
3. See how a computer can be used to generate magic squares.
4. Write a BASIC program to generate magic squares.

## 7.1 ORIGIN

For centuries the study and construction of magic squares have preoccupied both mathematicians and persons interested in the recreational aspects of mathematics. The first known example of a magic square is said to have been found on the back of a tortoise by the Emperor Yu in about 2200 B.C.E.!

This was called the *lo-shu* and appeared as a 3 × 3 array of numerals indicated by knots in strings as in the accompanying figure.

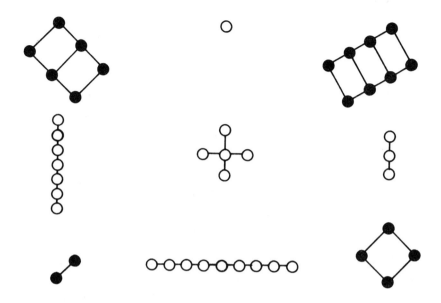

Black knots were used for even numbers and white ones for odd numbers. The sum along any row, column, or main diagonal is 15. The accompanying magic square depicts the *lo-shu* in decimal notation.

| 6 | 1 | 8 |
|---|---|---|
| 7 | 5 | 3 |
| 2 | 9 | 4 |

The magic square, still a common sight in China and India today, is found on buildings and in artistic designs, and every fortune teller makes use of it in the trade.

Because the strange properties of these squares were considered magical in medieval days, the squares served as talismans, protecting the wearer against many evils. An often reproduced magic square is the one in Albrecht Durer's

**Figure 7.1**  Melancholia

famous print Melancholia (see Figures 7.1 and 7.2), which shows how the digits were written during Durer's time. The middle numbers in the last line represent the year 1514, in which Durer's print was made. He probably started from these two numbers and found the remaining ones by trial and error.

In Europe the magic square has lost its value as a charm and talisman and become primarily a mathematical diversion for professional and amateur mathematicians alike.

## 7.2    WHAT IS A MAGIC SQUARE?

By a magic square we mean a square divided into $N^2$ cells in which the integers from 1 to $N^2$ are placed in such a manner that the sums of the rows, columns, and both diagonals are identical. The simplest magic square is one containing $3^2$ or 9 cells, with numbers from 1 to 9 inclusive, as shown in the following diagram. The sum of each row, column, and diagonal is 15.

| 8 | 1 | 6 |
|---|---|---|
| 3 | 5 | 7 |
| 4 | 9 | 2 |

The order of a magic square is the number of rows or columns. Hence, a magic square with $N^2$ cells has order $N$. If the square has $5^2$ or 25 cells it is of order 5.

We have seen that a magic square of order 3 has all rows and columns totaling 15 when numbers from 1 to 9 are used. A magic square of order 4 has rows and columns totaling 34 when all numbers from 1 to 16 are used. What about squares of order 5, 6, 7, or larger? What will the rows and columns of these total? The sum of the rows, columns, and main diagonals is called the magic constant, and is determined by the formula

$$\text{magic constant} = \frac{N(N^2 + 1)}{2}$$

For example, in the magic square with 9 cells the magic constant is determined as follows:

$$\text{magic constant} = \frac{3(3^2 + 1)}{2} = \frac{3(9 + 1)}{2} = \frac{30}{2} = 15$$

In an order 5 square, $N = 5$:

$$\frac{5(5^2 + 1)}{2} = \frac{5(25 + 1)}{2} = \frac{5(26)}{2} = \frac{130}{2} = 65$$

All rows, columns, and main diagonals total 65.
In a square with 25 rows and 25 columns, $N = 25$:

$$\frac{25(25^2 + 1)}{2} = \frac{25(625 + 1)}{2} = \frac{25(626)}{2} = \frac{15650}{2} = 7825$$

**Figure 7.2**  Close-up view of the Melancholia magic square.

Magic squares are classified as odd order or even order, with the even order squares further classified as singly even or doubly even. If the order, $N$, can be expressed as $2m + 1$, where $m$ is an integer, we have a magic square of odd order (3, 5, 7, 9, 11, ...); if $N$ can be written as $2(2m + 1)$, we have a singly even order (6, 10, 14, ...); and if $N$ is of the form $4m$, the square is doubly even (4, 8, 12, ...).

Magic squares of an odd order are usually constructed by methods that differ from those governing the construction of even order squares. The odd order square is easier to construct. Several methods for constructing magic squares are given below.

## 7.3 ODD ORDER MAGIC SQUARES

There is a general method for constructing magic squares of any odd order called the De la Loubere method. Let us demonstrate this method by constructing an order 3 magic square.

Always start by placing number 1 in the center cell of the top row. Now go right oblique upward and you will note that the next number, 2, comes above and outside this square in the dotted cell. In this case drop the 2 to the lowest cell in that column. Now go right oblique upwards again and you find the next number, 3, comes outside of the square to the right. In this case move the 3 to the extreme left cell in that row. Now note that the number 1 interferes with the next right oblique upward move; therefore, drop down one cell and place the 4 directly below the 3. Now you have clear sailing for 5 and 6. It is impossible to go right oblique from this corner (since there is no cell corresponding to this move) hence, drop down in the square as before, placing the 7 directly below the 6, and continue as before, placing the 8 and 9 in their proper locations as shown in the accompanying figure.

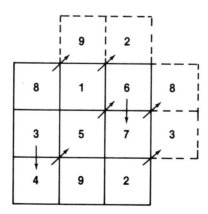

Practice constructing this magic square without the dotted lines as guides, then try the next odd order square, order 5 with 25 cells. Exactly the same principle is applied to this order 5 square, as a study of the following figure will reveal.

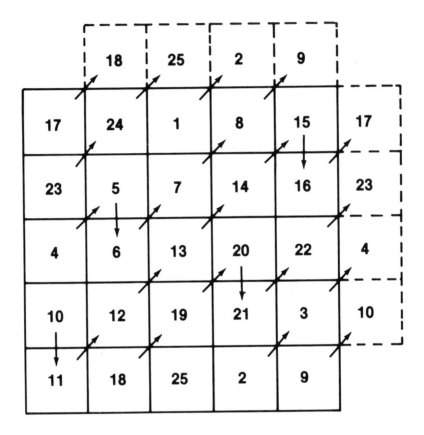

You will note that the first number is in the center of the top row, and the direction (indicated by the arrows) is always right oblique or down. You should practice making the order 5 square without using the dotted lines as guides. The same principle applies to the construction of all odd order magic squares. Practice constructing an order 7 or order 9 magic square.

The following BASIC program generates an order 5 magic square using the De la Loubere technique.

```
140    REM MAGIC SQUARE GENERATING PROGRAM
150    REM ORDER OF SQUARE IS N
160    LET N=5
170    LET K=1
180    LET N1=1
190    LET I=1
200    LET J=(N+1)/2
210    REM PLACE 1 IN CENTER CELL OF TOP ROW
220    LET M[I,J]=N1
230    LET N1=N1+1
240    REM IS MAGIC SQUARE GENERATION COMPLETE?
250    REM TRANSFER CONTROL TO 490 IF LAST
260    REM NUMBER HAS BEEN STORED IN TABLE M
270    IF N1>N↑2 THEN 490
280    REM IS K AN EVEN MULTIPLE OF N?
290    IF K<N THEN 350
300    REM RESET K TO 1
301    REM SET ROW INDEX TO NEXT ROW
310    LET K=1
320    LET I=I+1
330    GOTO 220
340    REM INCREASE K BY 1 AND MOVE RIGHT AND UP
350    LET K=K+1
360    LET I=I-1
370    LET J=J+1
380    REM DO NEW SUBSCRIPTS SPECIFY A
390    REM LOCATION OUTSIDE OF TABLE M
400    IF I <> 0 THEN 440
410    REM OUTSIDE OF SQUARE - RESET ROW IND TO N
420    LET I=N
430    GOTO 220
440    IF J <= N THEN 220
450    REM OUTSIDE OF SQUARE - RESET COL IND TO 1
460    LET J=1
470    GOTO 220
580    REM PRINT MAGIC SQUARE
490    FOR I=1 TO N
500    FOR J=1 TO N
510    PRINT M[I,J];
520    NEXT J
```

```
530    PRINT
540    PRINT
550    PRINT
560    NEXT I
570    END

RUN
```

| 17 | 24 | 1 | 8 | 15 |
|----|----|----|----|----|
| 23 | 5 | 7 | 14 | 16 |
| 4 | 6 | 13 | 20 | 22 |
| 10 | 12 | 19 | 21 | 3 |
| 11 | 18 | 25 | 2 | 9 |

The statement at line number 160 establishes the order of the square as 5; that is, the program will generate a 5 by 5 magic square. Starting values for $K$ and $N1$ are both 1. $K$ is a program counter used to determine multiples of $N$. $N1$ will vary from 1 to $N\uparrow2$, and each value of $N1$ is stored in array M.

The subscripts $I$ and $J$ are first set to 1 and $(N + 1)/2$, and the first row and the center cell of the first row are specified, respectively, when appended as subscripts to array M. After $N1$ is stored in cell $M(I, J)$, it is incremented by 1 and compared with the largest number to be entered, the square of $N$. If the value of $N1$ exceeds the square of $N$, the program causes an order $N$ magic square to be printed on the terminal. If the value of $N1$ does not exceed that highest value, the program continues with a check to determine whether $K$ is a multiple of $N$. If so, then $K$ is reset to 1, and the row subscript is set to specify the next row. If $K$ is not a multiple of $N$, then $K$ is increased by 1, and the subscripts $I$ and $J$ are updated to specify the next cell to the right oblique. If the new value of $I$ is less than 1, indicating that a location outside the top of the square is specified, the row subscript is set to address the last row of array M. If the new value of $J$ is greater than $N$, indicating that a location outside the right side of the square is specified, the column subscript is set to address the first column of array M. The new value of $N1$ is then stored in the new cell of $M(I, J)$. This process continues until $N1$ exceeds the square of $N$, when the magic square is complete.

This program generated a 5 by 5 magic square, but the procedure applies to any square having an odd number of cells in each row and column and requires only a change in the statement at line number 160, where the size of the square is specified. For example, if that statement is revised to read LET $N = 7$, the program will compute a 7 by 7 magic square. To compute a magic square of 81 cells that statement would be changed to LET $N = 9$. You must also change the PRINT statements to print the generated square in the shape of a square. A flowchart is shown in Figure 7.3.

Figure 7.4 is a computer generated order 21 magic square.

## 7.4  EVEN ORDER MAGIC SQUARES

Construction of magic squares of even order is more difficult than construction of magic squares of odd order. We shall discuss only the construction of squares of doubly even order here. The following example illustrates the construction of an order 4 magic square.

1.  In a blank 4 by 4 square, fill the main diagonal squares with X's.

|   |   |   |   |
|---|---|---|---|
| X |   |   | X |
|   | X | X |   |
|   | X | X |   |
| X |   |   | X |

2.  Start with the upper left square and move toward the right obeying the following rules.

    a.  If the cell is occupied by an X, do not number the cell.

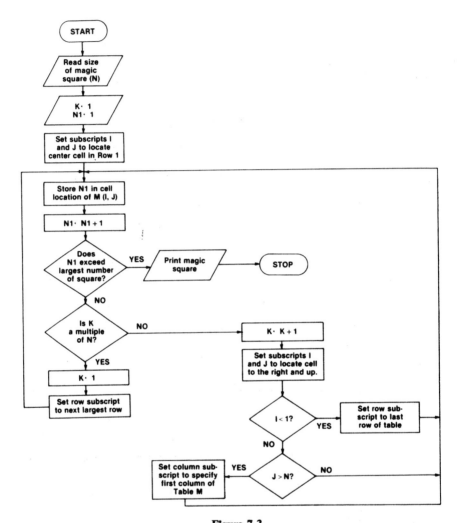

**Figure 7.3.**

| 233 | 256 | 279 | 302 | 325 | 348 | 371 | 394 | 417 | 440 | 1 | 24 | 47 | 70 | 93 | 116 | 139 | 162 | 185 | 208 | 231 |
|---|---|---|---|---|---|---|---|---|---|---|---|---|---|---|---|---|---|---|---|---|
| 255 | 278 | 301 | 324 | 347 | 370 | 393 | 416 | 439 | 21 | 23 | 46 | 69 | 92 | 115 | 138 | 161 | 184 | 207 | 230 | 232 |
| 277 | 300 | 323 | 346 | 369 | 392 | 415 | 438 | 20 | 22 | 45 | 68 | 91 | 114 | 137 | 160 | 183 | 206 | 229 | 252 | 254 |
| 299 | 322 | 345 | 368 | 391 | 414 | 437 | 19 | 42 | 44 | 67 | 90 | 113 | 136 | 159 | 182 | 205 | 228 | 251 | 253 | 276 |
| 321 | 344 | 367 | 390 | 413 | 436 | 18 | 41 | 43 | 66 | 89 | 112 | 135 | 158 | 181 | 204 | 227 | 250 | 273 | 275 | 298 |
| 343 | 366 | 389 | 412 | 435 | 17 | 40 | 63 | 65 | 88 | 111 | 134 | 157 | 180 | 203 | 226 | 249 | 272 | 274 | 297 | 320 |
| 365 | 388 | 411 | 434 | 16 | 39 | 62 | 64 | 87 | 110 | 133 | 156 | 179 | 202 | 225 | 248 | 271 | 294 | 296 | 319 | 342 |
| 387 | 410 | 433 | 15 | 38 | 61 | 84 | 86 | 109 | 132 | 155 | 178 | 201 | 224 | 247 | 270 | 293 | 295 | 318 | 341 | 364 |
| 409 | 432 | 14 | 37 | 60 | 83 | 85 | 108 | 131 | 154 | 177 | 200 | 223 | 246 | 269 | 292 | 315 | 317 | 340 | 363 | 386 |
| 431 | 13 | 36 | 59 | 82 | 105 | 107 | 130 | 153 | 176 | 199 | 222 | 245 | 268 | 291 | 314 | 316 | 339 | 362 | 385 | 408 |
| 12 | 35 | 58 | 81 | 104 | 106 | 129 | 152 | 175 | 198 | 221 | 244 | 267 | 290 | 313 | 336 | 338 | 361 | 384 | 407 | 430 |
| 34 | 57 | 80 | 103 | 126 | 128 | 151 | 174 | 197 | 220 | 243 | 266 | 289 | 312 | 335 | 337 | 360 | 383 | 406 | 429 | 11 |
| 56 | 79 | 102 | 125 | 127 | 150 | 173 | 196 | 219 | 242 | 265 | 288 | 311 | 334 | 357 | 359 | 382 | 405 | 428 | 10 | 33 |
| 78 | 101 | 124 | 147 | 149 | 172 | 195 | 218 | 241 | 264 | 287 | 310 | 333 | 356 | 358 | 381 | 404 | 427 | 9 | 32 | 55 |
| 100 | 123 | 146 | 148 | 171 | 194 | 217 | 240 | 263 | 286 | 309 | 332 | 355 | 378 | 380 | 403 | 426 | 8 | 31 | 54 | 77 |
| 122 | 145 | 168 | 170 | 193 | 216 | 239 | 262 | 285 | 308 | 331 | 354 | 377 | 379 | 402 | 425 | 7 | 30 | 53 | 76 | 99 |
| 144 | 167 | 169 | 192 | 215 | 238 | 261 | 284 | 307 | 330 | 353 | 376 | 399 | 401 | 424 | 6 | 29 | 52 | 75 | 98 | 121 |
| 166 | 189 | 191 | 214 | 237 | 260 | 283 | 306 | 329 | 352 | 375 | 398 | 400 | 423 | 5 | 28 | 51 | 74 | 97 | 120 | 143 |
| 188 | 190 | 213 | 236 | 259 | 282 | 305 | 328 | 351 | 374 | 397 | 420 | 422 | 4 | 27 | 50 | 73 | 96 | 119 | 142 | 165 |
| 210 | 212 | 235 | 258 | 281 | 304 | 327 | 350 | 373 | 396 | 419 | 421 | 3 | 26 | 49 | 72 | 95 | 118 | 141 | 164 | 187 |
| 211 | 234 | 257 | 280 | 303 | 326 | 349 | 372 | 395 | 418 | 441 | 2 | 25 | 48 | 71 | 94 | 117 | 140 | 163 | 186 | 209 |

**Figure 7.4**   A computer-generated order 21 magic square.

b.  If the cell is not occupied with an X, insert its number. Start with the number 1 in square 1 and increment the count by 1 each time a move is made. When the end of a row is reached, repeat the process in the next row.

3.  The first eight numbers would be placed in the square as follows:

| 16 |    |    | 13 |
|----|----|----|----|
|    | 11 | 10 |    |
|    | 7  | 6  |    |
| 4  |    |    | 1  |

4.  Now fill the cells containing an X. Start in the upper left square and obey the following rules:

a.  If the cell is occupied by an X, insert a number.

b.  If the cell is occupied by a number, do not number the cell. Start with the number 16 and decrease the count by 1 each time a move is made. When the end of a row is reached repeat the same process in the next row.

5.  The last eight numbers would be placed in the square in the following order:

| X | 2  | 3  | X  |
|---|----|----|----|
| 5 | X  | X  | 8  |
| 9 | X  | X  | 12 |
| X | 14 | 15 | X  |

6.   The completed magic square would appear as follows:

| 16 | 2  | 3  | 13 |
|----|----|----|----|
| 5  | 11 | 10 | 8  |
| 9  | 7  | 6  | 12 |
| 4  | 14 | 15 | 1  |

The following BASIC program uses this method to generate an order 4 magic square.

```
100  REM 4 BY 4 MAGIC SQUARE
110  LET N=4
120  REM STORE ZEROS IN ARRAY M
130  FOR I=1 TO N
140  FOR J=1 TO N
150  LET M[I,J]=0
160  NEXT J
170  NEXT I
180  REM STORE 999 IN EACH CELL OF DIAGONAL 1
190  FOR I=1 TO N
200  LET J=I
210  LET M[I,J]=999
220  NEXT I
230  REM STORE 999 IN EACH CELL OF DIAGONAL 2
240  FOR I=1 TO N
250  LET J=N-I+1
260  LET M[I,J]=999
270  NEXT I
```

```
280  REM FIRST PASS THROUGH ARRAY
290  LET K=1
300  FOR I=1 TO N
310  FOR J=1 TO N
320  IF M[I,J] <> 0 THEN 340
330  LET M[I,J]=K
340  LET K=K+1
350  NEXT J
360  NEXT I
370  REM SECOND PASS THROUGH ARRAY
380  LET K=N*N
390  FOR I=1 TO N
400  FOR J=1 TO N
410  IF M[I,J] <> 999 THEN 430
420  LET M[I,J]=K
430  LET K=K-1
440  NEXT J
450  NEXT I
460  REM PRINT MAGIC SQUARE
470  PRINT "4 BY 4 MAGIC SQUARE"
480  PRINT
490  FOR I=1 TO N
500  FOR J=1 TO N
510  PRINT M[I,J];
520  NEXT J
530  PRINT
540  PRINT
550  PRINT
560  NEXT I
570  END
```

RUN

4 BY 4 MAGIC SQUARE

```
 16    2     3    13

  5   11    10     8

  9    7     6    12

  4   14    15     1
```

The program stores 999 in each cell that lies on a diagonal and zeros in all other locations of array M. The program keeps a counter $K$ that varies from 1 to 16 on the first pass through the array and from 16 to 1 on the second pass. The counter value is adjusted as the program steps through the array.

If a cell of array M contains a zero on the first pass through the array, the counter value is placed in this cell. The value of the cell is not changed if it appears on a diagonal. On the second pass through the array, the counter value is stored in the cell if the cell contains 999. All other elements of the array remain unchanged. The calculated order 4 magic square is printed before the program stops.

The next magic square of doubly even order is of order 8. It can be constructed in a manner similar to the method used previously for the order 4 magic square. First, consider that the order 8 magic square is subdivided into four squares of order 4, and then visualize diagonals filled with X's drawn in each of these order four squares. Figure 7.5 illustrates how the diagonals are marked by placing X's on all diagonals of the order 4 squares. The procedure applied to the generation of order 4 squares also applies to the generation of order 8 squares, but the number range includes the numbers 1 through 64. You should construct an order 8 magic square using the procedure shown in Figure 7.5

Figure 7.6 illustrates an order 20 magic square generated by this procedure. This 400-cell square is composed of twenty-five order 4 squares.

## 7.5 MAGIC SQUARES STARTING WITH NUMBERS OTHER THAN ONE

The magic squares previously shown started with the number 1. However, magic squares may start with any number.

The following order 3 magic square starts with 4.

| 11 | 4 | 9 |
|----|----|----|
| 6 | 8 | 10 |
| 7 | 12 | 5 |

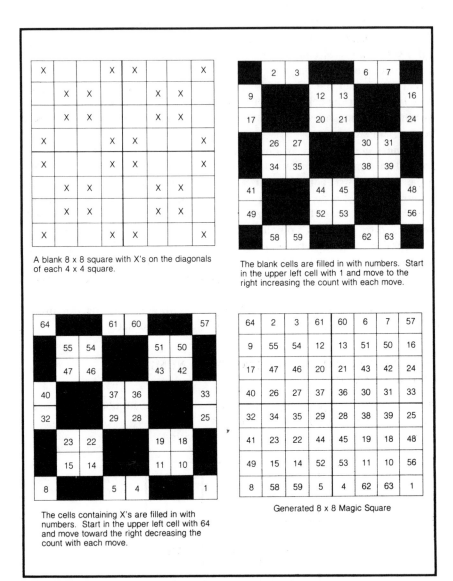

A blank 8 x 8 square with X's on the diagonals of each 4 x 4 square.

The blank cells are filled in with numbers. Start in the upper left cell with 1 and move to the right increasing the count with each move.

The cells containing X's are filled in with numbers. Start in the upper left cell with 64 and move toward the right decreasing the count with each move.

Generated 8 x 8 Magic Square

**Figure 7.5**   Procedure for generating an order 8 magic square.

| | | | | | | | | | | | | | | | | | | | |
|---|---|---|---|---|---|---|---|---|---|---|---|---|---|---|---|---|---|---|---|
| 400 | 2 | 3 | 397 | 396 | 6 | 7 | 393 | 392 | 10 | 11 | 389 | 398 | 14 | 15 | 385 | 384 | 18 | 19 | 381 |
| 21 | 379 | 378 | 24 | 25 | 375 | 374 | 28 | 29 | 371 | 370 | 32 | 33 | 367 | 366 | 36 | 37 | 363 | 362 | 40 |
| 41 | 359 | 358 | 44 | 45 | 355 | 354 | 48 | 49 | 351 | 350 | 52 | 53 | 347 | 346 | 56 | 57 | 343 | 342 | 60 |
| 340 | 62 | 63 | 337 | 336 | 66 | 67 | 333 | 332 | 70 | 71 | 329 | 328 | 74 | 75 | 325 | 324 | 78 | 79 | 321 |
| 320 | 82 | 83 | 317 | 316 | 86 | 87 | 313 | 312 | 90 | 91 | 309 | 308 | 94 | 95 | 305 | 304 | 98 | 99 | 301 |
| 101 | 299 | 298 | 104 | 105 | 295 | 294 | 108 | 109 | 291 | 290 | 112 | 113 | 287 | 286 | 116 | 117 | 283 | 282 | 120 |
| 121 | 279 | 278 | 124 | 125 | 275 | 274 | 128 | 129 | 271 | 270 | 132 | 133 | 267 | 266 | 136 | 137 | 263 | 262 | 140 |
| 268 | 142 | 143 | 257 | 256 | 146 | 147 | 253 | 252 | 150 | 151 | 249 | 248 | 154 | 155 | 245 | 244 | 158 | 159 | 241 |
| 240 | 162 | 163 | 237 | 236 | 166 | 167 | 233 | 232 | 170 | 171 | 229 | 228 | 174 | 175 | 225 | 224 | 178 | 179 | 221 |
| 181 | 219 | 218 | 184 | 185 | 215 | 214 | 188 | 189 | 211 | 210 | 192 | 193 | 207 | 206 | 196 | 197 | 203 | 202 | 200 |
| 201 | 199 | 198 | 204 | 205 | 195 | 194 | 208 | 209 | 191 | 190 | 212 | 213 | 187 | 186 | 216 | 217 | 183 | 182 | 220 |
| 180 | 222 | 223 | 177 | 176 | 226 | 227 | 173 | 172 | 230 | 231 | 169 | 168 | 234 | 235 | 165 | 164 | 238 | 239 | 161 |
| 160 | 242 | 243 | 157 | 156 | 246 | 247 | 153 | 152 | 250 | 151 | 149 | 148 | 254 | 255 | 145 | 144 | 258 | 259 | 141 |
| 261 | 139 | 138 | 264 | 265 | 135 | 134 | 268 | 269 | 131 | 130 | 272 | 273 | 127 | 126 | 276 | 277 | 123 | 122 | 280 |
| 281 | 119 | 118 | 284 | 285 | 115 | 114 | 288 | 289 | 111 | 110 | 292 | 293 | 107 | 106 | 296 | 297 | 103 | 102 | 300 |
| 100 | 302 | 303 | 97 | 96 | 306 | 307 | 93 | 92 | 310 | 311 | 89 | 88 | 314 | 315 | 85 | 84 | 318 | 319 | 81 |
| 80 | 322 | 323 | 77 | 76 | 326 | 327 | 73 | 72 | 330 | 331 | 69 | 68 | 334 | 335 | 65 | 64 | 338 | 339 | 61 |
| 341 | 59 | 58 | 344 | 345 | 55 | 54 | 348 | 349 | 51 | 50 | 352 | 353 | 47 | 46 | 356 | 357 | 43 | 42 | 360 |
| 361 | 39 | 38 | 364 | 365 | 35 | 34 | 368 | 369 | 31 | 30 | 372 | 373 | 27 | 26 | 376 | 377 | 23 | 22 | 380 |
| 20 | 382 | 383 | 17 | 16 | 386 | 387 | 13 | 12 | 390 | 391 | 9 | 8 | 394 | 395 | 5 | 4 | 398 | 399 | 1 |

**Figure 7.6** A computer-generated order 20 magic square.

The magic constant of this square is 24 and is computed by the formula

$$\text{magic constant} = \frac{N^3 + N}{2} + N(P - 1) = \frac{3^3 + 3}{2} + 3(4 - 1) = 24$$

where $N$ is the order of the square and $P$ is the starting number.

The following illustration shows an order 4 square that starts with 4. The magic constant of this square is 46.

| 4 | 15 | 10 | 17 |
| 11 | 16 | 5 | 14 |
| 13 | 6 | 19 | 8 |
| 18 | 9 | 12 | 7 |

The following BASIC program will generate an odd order magic square starting with any number.

```
100  REM MAGIC SQUARE - STARTING WITH ANY NO.
110  DIM M[25,25]
120  PRINT "TYPE SIZE OF SQUARE";
130  INPUT N
140  PRINT "TYPE STARTING NUMBER";
150  INPUT Y
155  LET S=Y
160  PRINT N;"BY";N;"MAGIC SQUARE STARTING"
161  PRINT "WITH THE NUMBER";S
170  PRINT
180  LET K=1
190  LET I=1
200  LET J=(N+1)/2
210  REM PLACE THE FIRST NUMBER IN CENTER
211  REM CELL OF THE TOP ROW
220  LET M[I,J]=S
230  LET S=S+1
240  REM IS MAGIC SQUARE COMPLETE?
250  REM HAS LAST NO. BEEN STORED IN ARRAY M?
```

```
260   REM IF YES - PRINT MAGIC SQUARE
270   IF S>N↑2+Y-1 THEN 490
280   REM IS K AN EVEN MULTIPLE OF N
290   IF K<N THEN 350
300   REM RESET K TO 1 AND SET ROW INDEX
301   REM TO INDICATE THE NEXT ROW
310   LET K=1
320   LET I=I+1
330   GOTO 220
340   REM MOVE POSITION TO THE RIGHT AND UP,AND
341   REM INCREASE K BY 1
350   LET I=I-1
360   LET J=J+1
370   LET K=K+1
380   REM DO SUBSCRIPTS NOW SPECIFY A
390   REM LOCATION OUTSIDE OF ARRAY M
400   IF I <> 0 THEN 440
410   REM OUT OF SQUARE - RESET ROW INDEX TO 1
420   LET I=N
430   GOTO 220
440   IF J <= N THEN 220
450   REM OUT OF SQUARE - RESET COL INDEX TO 1
460   LET J=1
470   GOTO 220
480   REM PRINT MAGIC SQUARE
490   FOR I=1 TO N
500   FOR J=1 TO N
510   PRINT M[I,J];
520   NEXT J
530   PRINT
540   PRINT
550   PRINT
560   NEXT I
570   END

RUN

TYPE SIZE OF SQUARE?7
TYPE STARTING NUMBER?784
  7    BY 7    MAGIC SQUARE STARTING
WITH THE NUMBER 784
```

| | | | | | | |
|---|---|---|---|---|---|---|
| 813 | 822 | 831 | 784 | 793 | 802 | 811 |
| 821 | 830 | 790 | 792 | 801 | 810 | 812 |
| 829 | 789 | 791 | 800 | 809 | 818 | 820 |
| 788 | 797 | 799 | 808 | 817 | 819 | 828 |
| 796 | 798 | 807 | 816 | 825 | 827 | 787 |
| 804 | 806 | 815 | 824 | 826 | 786 | 795 |
| 805 | 814 | 823 | 832 | 785 | 794 | 803 |

Input to the program is the order of the square and the starting number. In the example shown, 7 is the order and 784 is the starting number.

After receiving the input information the program causes the following heading to be typed:

### 5 BY 5 MAGIC SQUARE — STARTING WITH 784

The program sets the subscripts $I$ and $J$ to locate the middle cell in the first row of array M. The starting value is stored in this location. The starting value is increased by 1, and a check is made to see if the program has stored $N^2$ values in array M. If all values have been stored, the program will output the magic square. If the program is not through calculating, another check is made to determine whether $K$ is an even multiple of $N$; and if so, $K$ is reset to 1 and the row indicator $I$ is advanced to the next row. If $K$ is not a multiple of $N$, $K$ is advanced by 1, and the subscripts $I$ and $J$ are set to address the next cell of array M that is to the right and up. If the new value of $J$ indicates a cell location outside the right side of array M, the column indicator $J$ is reset to the first column of the array. If the new value of $I$ is less than 1, $I$ is reset to $N$. The program then stores the correct number in the array M and the program continues until $K$ exceeds the maximum value to be stored in the array.

## 7.6   MULTIPLICATION MAGIC SQUARE

An order 3 multiplication magic square is shown below. The magic constant is 216 and is obtained by multiplying together the three numbers in any column, row, or diagonal.

| 18 | 1 | 12 |
|----|----|----|
| 4 | 6 | 9 |
| 3 | 36 | 2 |

A method based on the De la Loubere odd order construction method may be used to generate multiplication magic squares of odd order. The construction of a 5 by 5 multiplication square is used to illustrate the method.

1.  Place the number 1 in the center cell of the first row in a blank 5 by 5 square.
2.  Move in an oblique direction, one square above and to the right. This movement results in leaving the top of the box. It is necessary to place the next number, twice the last number, or 2, in a cell at the bottom of the column in which you wanted to place the number.
3.  Now move diagonally to the right again and put the number twice the last, or 4, in the next cell location.
4.  If you continue diagonally to the right you leave the cell on the right side. When this occurs, you must go to the extreme left of the row in which you wanted to place the next number. After crossing over to the left side of the square, place the number twice that of the last into the appropriate cell.
5.  Now, again, go up diagonally to the right and place the next number determined by doubling the last number.

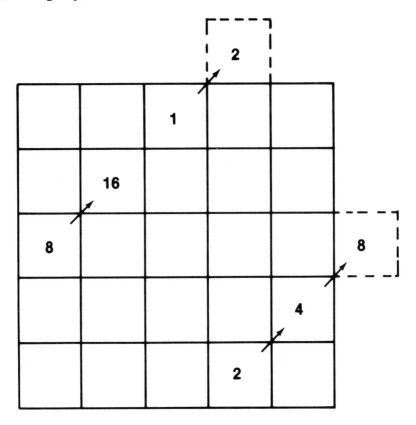

This completes the first group of five numbers of an order 5 square. The next group of five numbers starts with 3, the next group with 9, the next group with 27, and the last group with 81. The reader should note that the starting numbers are all powers of 3:

$$3^0 = 1$$
$$3^1 = 3$$
$$3^2 = 9$$
$$3^3 = 27$$
$$3^4 = 81$$

6.  Since this is a 5 by 5 square, you must move down one cell to place the next group of five numbers. The number placed in this cell is the starting number of the second group of five numbers. In the case of a 3 by 3 square, you would move down when you completed a group of 3.

7.  Move up diagonally to the right and place a number into each cell you enter. If you leave the top of the box, move to the bottom of the column in which you wanted to place the number. If you move outside the box on the right side, move across to the opposite side. The next number is always determined by doubling the last number. When you finish the second group of five numbers, the square should appear as shown.

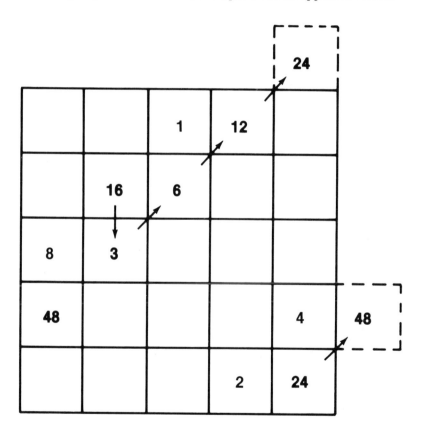

8.  The following square would result after the third group of five numbers have been placed. The starting number is 9.

| | | 1 | 12 | 144 |
|---|---|---|---|---|
| | 16 | 6 | 72 | |
| 8 | 3 | 36 | | |
| 48 | 18 | | | 4 |
| 9 | | | 2 | 24 |

9.  The fourth group of five numbers would be placed in a similar manner.
    The starting number is 27.

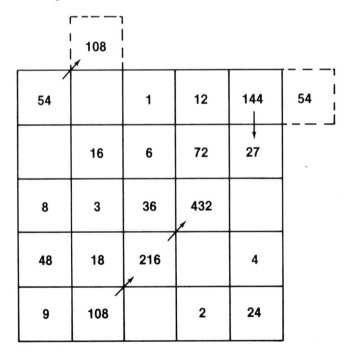

10. When the twenty-fifth number is placed in the cell opposite the starting cell, the final magic square will appear as shown.

| 54 | 648 | 1 | 12 | 144 |
|---|---|---|---|---|
| 324 | 16 | 6 | 72 | 27 |
| 8 | 3 | 36 | 432 | 162 |
| 48 | 18 | 216 | 81 | 4 |
| 9 | 108 | 1296 | 2 | 24 |

The magic constant of this order 5 square is 60 466 176.

## 7.7 GEOMETRIC MAGIC SQUARE

A geometric magic square is an array of numbers in which the product of the numbers in every column, row, and main diagonal is the same and each number of the square is represented by a base value and an exponent. The base value remains the same in all the positions of the square, and the exponent values are the numbers in an ordinary odd order magic square. For example, an order 3 geometric magic square with a base of 2 would appear as follows:

| | | |
|---|---|---|
| $2^8$ | $2^1$ | $2^6$ |
| $2^3$ | $2^5$ | $2^7$ |
| $2^4$ | $2^9$ | $2^2$ |

or

| | | |
|---|---|---|
| 256 | 2 | 64 |
| 8 | 32 | 128 |
| 16 | 512 | 4 |

Geometric magic
square using base
and exponent values

Geometric magic
square using
integer values

The following BASIC program generates an order 3 geometric magic square with a base of 2.

```
10 REM GEOMETRIC MAGIC SQUARE
20 REM PROGRAM GENERATES AN ODD ORDER
30 REM GEOMETRIC MAGIC SQUARE OF SIZE N BY N
40 PRINT "SIZE OF SQUARE TO BE GENERATED IS";
50 INPUT N
60 PRINT "BASE OF SQUARE IS";
70 INPUT B
80 LET K=1
90 LET A=B
100   LET L=1
110   LET I=1
120   LET J=(N+1)/2
130   LET G[I,J]=A
140   LET L=L+1
150   LET A=B↑L
160   IF A>(B↑(N↑2)) THEN 295
170   IF K<N THEN 210
180   LET K=1
```

```
190   LET I=I+1
200   GOTO 130
210   LET K=K+1
220   LET I=I-1
230   LET J=J+1
240   IF I <> 0 THEN 270
250   LET I=N
260   GOTO 130
270   IF J <= N THEN 130
280   LET J=1
290   GOTO 130
295   PRINT
296   PRINT
300   PRINT N;"BY";N;"GEOMETRIC MAGIC SQUARE"
310   PRINT
320   FOR I=1 TO N
330   FOR J=1 TO N
340   PRINT G[I,J];
350   NEXT J
360   PRINT
370   PRINT
380   PRINT
390   NEXT I
400   END

RUN

SIZE OF SQUARE TO BE GENERATED IS?3
BASE OF SQUARE IS?2


 3    BY 3  GEOMETRIC MAGIC SQUARE

256    2    64


 8    32   128


16    512  4
```

Shown below are order 3 geometric magic squares with bases of 4 and 5.

| 65536 | 4 | 4096 |
|---|---|---|
| 64 | 1024 | 16384 |
| 256 | 262144 | 16 |

Base 4

| 390625 | 5 | 15625 |
|---|---|---|
| 125 | 3125 | 78125 |
| 625 | 1953125 | 25 |

Base 5

## 7.8  OTHER INTERESTING MAGIC SQUARES

A square that is magic for both addition and multiplication is shown in Figure 7.7. The magic constant for addition is 1200, while the multiplication constant is 1 619 541 385 529 760 000.

| 17 | 171 | 126 | 54 | 230 | 100 | 93 | 264 | 145 |
|---|---|---|---|---|---|---|---|---|
| 124 | 66 | 290 | 85 | 57 | 168 | 162 | 23 | 225 |
| 216 | 115 | 75 | 279 | 198 | 29 | 170 | 76 | 42 |
| 261 | 186 | 33 | 210 | 68 | 38 | 200 | 135 | 69 |
| 50 | 270 | 92 | 87 | 248 | 165 | 21 | 153 | 114 |
| 105 | 51 | 152 | 150 | 27 | 207 | 116 | 62 | 330 |
| 138 | 25 | 243 | 132 | 58 | 310 | 95 | 63 | 136 |
| 190 | 84 | 34 | 184 | 125 | 81 | 297 | 174 | 31 |
| 99 | 232 | 155 | 19 | 189 | 102 | 46 | 250 | 108 |

**Figure 7.7**  A square that is completely magic for both addition and multiplication.

| 184 | 217 | 170 | 75 | 188 | 219 | 172 | 77 | 228 | 37 | 86 | 21 | 230 | 39 | 88 | 25 |
|---|---|---|---|---|---|---|---|---|---|---|---|---|---|---|---|
| 169 | 74 | 185 | 218 | 171 | 76 | 189 | 220 | 85 | 20 | 229 | 38 | 87 | 24 | 231 | 40 |
| 216 | 183 | 68 | 167 | 222 | 187 | 78 | 173 | 36 | 227 | 22 | 83 | 42 | 237 | 26 | 89 |
| 73 | 168 | 215 | 186 | 67 | 174 | 221 | 190 | 19 | 84 | 35 | 238 | 23 | 90 | 41 | 232 |
| 182 | 213 | 166 | 69 | 178 | 223 | 176 | 79 | 226 | 33 | 82 | 31 | 236 | 43 | 92 | 27 |
| 165 | 72 | 179 | 214 | 175 | 66 | 191 | 224 | 81 | 18 | 239 | 34 | 91 | 30 | 233 | 44 |
| 212 | 181 | 70 | 163 | 210 | 177 | 80 | 161 | 48 | 225 | 32 | 85 | 46 | 235 | 28 | 93 |
| 71 | 164 | 211 | 180 | 65 | 162 | 209 | 192 | 17 | 96 | 47 | 240 | 29 | 94 | 45 | 234 |
| 202 | 13 | 126 | 61 | 208 | 15 | 128 | 49 | 160 | 241 | 130 | 97 | 148 | 243 | 132 | 103 |
| 125 | 60 | 203 | 14 | 127 | 64 | 193 | 16 | 129 | 112 | 145 | 242 | 131 | 102 | 149 | 244 |
| 12 | 201 | 62 | 123 | 2 | 207 | 50 | 113 | 256 | 159 | 98 | 143 | 246 | 147 | 104 | 133 |
| 59 | 124 | 11 | 204 | 63 | 114 | 1 | 194 | 111 | 144 | 255 | 146 | 101 | 134 | 245 | 150 |
| 200 | 9 | 122 | 55 | 206 | 3 | 116 | 51 | 158 | 253 | 142 | 99 | 154 | 247 | 136 | 105 |
| 121 | 58 | 205 | 10 | 115 | 54 | 195 | 4 | 141 | 110 | 155 | 254 | 135 | 100 | 151 | 248 |
| 8 | 199 | 56 | 119 | 6 | 197 | 52 | 117 | 252 | 157 | 108 | 139 | 250 | 153 | 106 | 137 |
| 57 | 120 | 7 | 198 | 53 | 118 | 5 | 196 | 109 | 140 | 251 | 156 | 107 | 138 | 249 | 152 |

**Figure 7.8** An order 16 magic square constructed with moves of the knight.

Another interesting square is shown in Figure 7.8. The magic constant is 2056, but the unusual nature in this square lies in its method of construction. If the numbers are followed consecutively, it is found that the moves from one to the next are the moves of the knight on a chessboard.

The following magic square is made up entirely of prime numbers.

| 41 | 113 | 59 |
|---|---|---|
| 89 | 71 | 53 |
| 83 | 29 | 101 |

## 7.9   THE MAGIC TRIANGLE

The triangle in the following diagram is a right triangle. C is the square on the hypotenuse and A and B are the squares on the two equal sides.

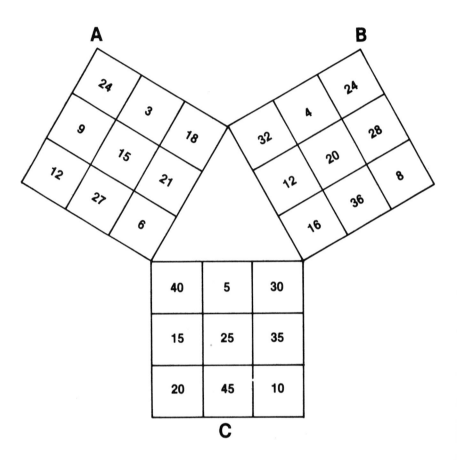

If you examine the diagram you will see that A, B, and C are each magic squares. The square of any cell of C is equal to the sum of the squares of the corresponding cells in A and B. For example, $30^2$ is equal to $18^2$ plus $24^2$.

Likewise, the square of the sum of any two or more cells or any diagonal or horizontal row or any perpendicular column in C is equal to the square of the sum of the corresponding two or more cells, rows, columns, or diagonals of A plus B.

Also note that the square of the total of the cells in C equals the square of the total of the cells in A plus the square of the total of the cells in B.

## Review Exercises

1. Show that the sum of the numbers of the second and fourth columns of the Melancholia magic square is equal to the sum of the numbers of the first and third rows.

2. What is the magic constant of the magic square of order (a) 6? (b) 7? (c) 15? (d) 21? Give an example of (e) a singly even order magic square; (f) a doubly even order magic square.

3. Fill in the blank square with numbers determined by the given relationships. The square will be magic if the numbers are correct.

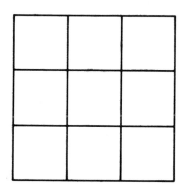

|   |   |   |
|---|---|---|
| a | d | g |
| b | e | h |
| c | f | i |

$a = 2^3$

$b = ?$

$c = ?$

$d = ?$

$e = \dfrac{g + h + i}{3}$

$f = ?$

$g = 15 - h - i$

$h =$ largest digit in the octal number system

$i =$ only even prime

4.  Complete the magic square.

| 16 |    | 3  | 13 |
|----|----|----|----|
| 5  | 11 | 10 |    |
|    |    | 6  |    |
| 4  | 14 |    |    |

5.  The following diagram illustrates a form that may be used to generate one of the eight possible arrangements of an order 3 magic square.

| $a + b$     | $a - (b + c)$ | $a + c$       |
|-------------|---------------|---------------|
| $a - (b - c)$ | $a$         | $a + (b - c)$ |
| $a - c$     | $a + (b + c)$ | $a - b$       |

Hand calculate a magic square using the values $a = 5$, $b = 1$, and $c = 3$. Write a BASIC program that uses this method to generate an order 3 magic square.

6. Hand construct a magic square of order 7.

7. Hand construct a magic square of order 9.

8. Modify the BASIC program given in Section 7.3 to generate an order 7 magic square.

9. Modify the BASIC program given in Section 7.3 to generate an order 11 magic square.

10. Write a BASIC program to prove that the order 21 magic square shown in Figure 7.4 is indeed a magic square.

11. Look up the Agrippa method of constructing magic squares of odd order in *Game Playing with Computers* by Donald D. Spencer (Hayden Book Company), and write a BASIC program to generate an order 5 magic square using this method.

12. Write a BASIC program to determine whether the following number arrangement is a magic square.

| 35 | 1  | 6  | 26 | 19 | 24 |
| 3  | 32 | 7  | 21 | 23 | 25 |
| 31 | 9  | 2  | 22 | 27 | 30 |
| 8  | 28 | 33 | 17 | 10 | 15 |
| 30 | 5  | 34 | 12 | 14 | 16 |
| 4  | 36 | 29 | 13 | 18 | 11 |

13. Hand construct a magic square of order 12. (Think of it as being subdivided into nine magic squares of order 4. Then proceed as we did in the construction of an order 8 magic square).

14. Draw a flowchart that could have been used to produce the BASIC program given in Section 7.4.

15. Write a BASIC program to generate an order 8 magic square.

16. Write a BASIC program to generate an order 12 magic square.

17. Write a BASIC program to generate an order 3 magic square starting with 7.

18. Write a BASIC program to generate an order 5 magic square that begins with 22.

19. Write a BASIC program to generate an order 7 magic square that begins with 43.

20. Draw a flowchart and write a BASIC program to generate the order 5 multiplication magic square discussed in Section 7.6.

21. Modify the BASIC program in Section 7.7 to generate an order 3 geometric magic square with base 4.

22. Write a BASIC program to generate the three order 3 magic squares that make up the magic triangle shown in Section 7.9.

# Chapter 8

# COMPUTERS AND THEIR
# NUMERATION SYSTEMS

**Preview**

Until the advent of computers, the decimal system reigned supreme in all fields of numerical calculations. The interest expressed in other systems was mainly historical and cultural. Only a few isolated problems in mathematics could best be stated by numeration systems other than the decimal system; one of the favorite examples in number theory was the game of Nim, which is discussed in Section 8.3.

Since computers operate on binary numbers, and computer programs sometimes contain octal or hexadecimal numbers, it is appropriate for students of number theory to understand these numeration sytems.

After completing this chapter, you should be able to:

1. Define number base or radix, digit, radix point, and bit.
2. Understand why computers use binary numbers rather than the common decimal numbers.
3. Represent binary, octal, and hexadecimal numbers in decimal system notation.
4. Play the ancient game of Nim.

## 8.1 INTRODUCTION

Most modern computers use the binary digit as a basic unit of information because the two digits of the binary system are easier to represent and use electronically than are the ten Arabic numerals of the decimal system. However, since people normally use the decimal system, it is necessary to provide a method for humans to communicate with the computer. This usually results in the computer using an internal code of binary notation, whereas input and output information to and from the computer is either decimal, octal, or hexadecimal notation. Of course, this involves a translation of the input and output information.

In most cases, it will be sufficient for you to think and use decimal values. In some cases, however, a knowledge of binary, octal, and hexadecimal representation is necessary.

As you will see later in the chapter, the fundamental concept underlying every number system is that of combining individuals into groups, these groups into larger groups, and so on. The number of individuals in a group, and the number of groups in a larger group, is unique for each number system. This number is known as the base or radix of that system. The base of a system determines the number that will represent a given quantity in that system. The number representation will be unique for a particular system, but will generally differ from that of the same quantity in another system. For example, we can write

$$1000001_2 = 101_8 = 65_{10} = 41_{16}$$

since these numbers all refer to the same quantity. The subscript attached to each number denotes the base used for that number. The numbers would be described as a binary number, an octal number, a decimal number, and a hexadecimal number. In general, we shall omit the subscripts on decimal (base 10) numbers.

A dot or period, usually called a decimal point, will be given that designation only when used with a decimal system. It is often called the radix point, although names such as binary point, hexadecimal point, and octal point will be used when those particular number systems are being discussed.

In any number system, a single character is called a digit. A binary digit is generally called a bit.

## 8.2  BINARY NUMBERS

In the binary system the base is 2, and digits can have one of two values, 0 or 1. Binary numbers are the common internal system for digital computation due to the relative simplicity of recording, storing, and recognizing variables of only two values. The value of a binary number is computed by multiplying the value of each digit by the corresponding power of 2 and summing all the products. The positional weights of a binary number (powers of 2) can be represented as shown in Figure 8.1.

It can be seen from Figure 8.1 that the powers of 2 are of ascending order to the left of the binary point and of descending order to the right of the binary

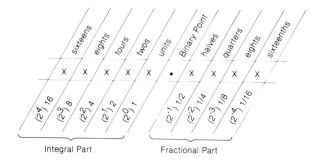

**Figure 8.1** The positional weights of a binary number (powers of 2).

point. The positive and negative powers of 2 and their decimal equivalents are shown in Table 8.1.

The presence of a 1 in a digit position of a binary number indicates that the corresponding power of 2 is used in determining the value of the binary number. A 0 in a digit position indicates that the corresponding power of 2 is absent from the binary number. For example, the binary number 101100 may be expressed as:

$$101100 = 1 \times 2^5 + 0 \times 2^4 + 1 \times 2^3 + 1 \times 2^2 + 0 \times 2^1 + 0 \times 2^0$$
$$= 1 \times 32 + 0 \times 16 + 1 \times 8 + 1 \times 4 + 0 \times 2 + 0 \times 1$$
$$= \quad 32 + \quad 0 + \quad 8 + \quad 4 + \quad 0 + \quad 0$$
$$= \quad 44$$

The binary number 101100 is equivalent to the decimal number 44.

## 8.3 A BINARY GAME

The game of Nim is usually played with three piles of counters between two players, one of which could be a computer.

Each player draws alternately one or more counters, up to the whole of one pile if he chooses. He can take from only one pile at a time, but for each turn can choose the pile from which to take. The player to take the last counter is the winner.

Like so many games, the outcome, if there are no mistakes, is predictable. For example, once one pile has been exhausted, the player drawing next can

**Table 8.1**  The Powers of 2 and Their Decimal Equivalents.

| Positive Power of 2 | Decimal Equivalent | Negative Power of 2 | Decimal Equivalent |
|---|---|---|---|
| $2^9$ | 512 | $2^{-1}$ | 1/2 |
| $2^8$ | 256 | $2^{-2}$ | 1/4 |
| $2^7$ | 128 | $2^{-3}$ | 1/8 |
| $2^6$ | 64 | $2^{-4}$ | 1/16 |
| $2^5$ | 32 | $2^{-5}$ | 1/32 |
| $2^4$ | 16 | $2^{-6}$ | 1/64 |
| $2^3$ | 8 | $2^{-7}$ | 1/128 |
| $2^2$ | 4 | $2^{-8}$ | 1/256 |
| $2^1$ | 2 | $2^{-9}$ | 1/512 |
| $2^0$ | 1 | | |

take enough counters from the biggest pile to leave the same number in each remaining pile. He must then win if he copies his opponent's moves exactly, always leaving two piles the same; this is a winning position.

This is not the only winning position, however. Surprisingly, a winning position can best be expressed in binary terms. Let us now consider a game with three piles of counters. The numbers contained in each pile are first written in the binary scale, and the three numbers set out as an addition sum. Each column is then added separately with the result expressed as decimal numbers. The total of each column will be 0, 1, 2, or 3. If at the outset, any of the totals is either 1 or 3, then the player drawing first can put himself in a winning position by taking sufficient counters from one pile to make the totals for all columns even, either 0 or 2. If at the outset all the totals are even, then the first player is in a losing position.

An example will make all this clearer.

If the piles begin at 12, 7, and 13, these will be written in the binary notation as:

| | |
|---:|---|
| 12 | 1100 |
| 7 | 111 |
| 13 | 1101 |
| Column total (in decimal scale) | 2312 |

There are odd totals in two columns. This can be converted to a winning position by the first player subtracting sufficient counters from one pile to make the column totals all even. There are several possibilities. For example, six counters from the pile of seven would make the column totals 2202, or two counters from the pile of twelve would make the totals 2222. An example of a losing position is a game in which the piles begin at 6, 9, and 15. The column total of the binary equivalents is 2222, and there is nothing that the first player can do that will keep all four columns even. Any number of counters removed will leave a 1 or 3 in the totals, enabling the second player to convert it to a winning position. The only possibility of gaining the initiative is by subtracting such a number that will make the task of deciding what can be taken to leave the column totals cell even as difficult as possible, then relying on the inferior skill of the opponent.

## 8.4  DECIMAL NUMBERS

The familiar decimal system is based on a radix of 10, and is composed of the digits 0, 1, 2, 3, 4, 5, 6, 7, 8, and 9. Each position of a decimal number has a weight of some power of 10. For example, the decimal number 8136 is described as follows:

$$
\begin{array}{llll}
8 & 1 & 3 & 6
\end{array}
$$

$$
\begin{aligned}
6 \times 10^0 &= 6 \times & 1 &= & 6 \\
3 \times 10^1 &= 3 \times & 10 &= & 30 \\
1 \times 10^2 &= 1 \times & 100 &= & 100 \\
8 \times 10^3 &= 8 \times & 1000 &= & \underline{8000} \\
& & & & 8136
\end{aligned}
$$

It can readily be seen in the foregoing example that each digit position has a value equal to the product of the digit appearing in the position and a corresponding power of 10.

The positional weights of a decimal number (powers of 10) can be represented as shown in Figure 8.2.

Any decimal number can be considered as the sum of the products obtained by multiplying the digits by the corresponding powers of the base. As an illustration of positional notation of a decimal number, consider the following decimal values:

$$
\begin{aligned}
6321 &= 6 \times 10^3 + 3 \times 10^2 + 2 \times 10^1 + 1 \times 10^0 \\
&= 6 \times 1000 + 3 \times 100 + 2 \times 10 + 1 \times 1 \\
&= 6000 \qquad + 300 \qquad + 20 \qquad + 1
\end{aligned}
$$

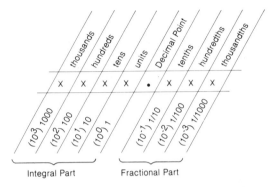

**Figure 8.2**   The positional weights of a decimal number (powers of 10).

## 8.5   HEXADECIMAL NUMBERS

The hexadecimal system is based on a radix of 16 and uses 16 digits. The digits 0 through 9 are used in the usual sense, and the other six digits are represented by the symbols A, B, C, D, E, and F.

The positional weights of a hexadecimal number (powers of 16) can be represented as shown in Figure 8.3.

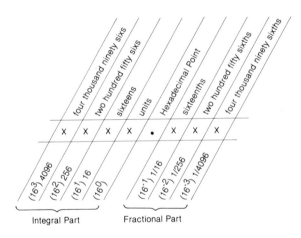

**Figure 8.3**   The positional weights of a hexadecimal number (powers of 16).

A decimal value of a hexadecimal number is determined by multiplying the value of each digit by the corresponding power of 16 and summing all the products. For example, the equivalent decimal value of the hexadecimal number 95.4 may be determined in the following manner:

$$9 \quad 5. \quad 4$$

$$4 \times 16^{-1} = 4 \times \frac{1}{16} = \frac{4}{16} = \quad .25$$

$$5 \times 16^0 = 5 \times 1 \quad = \quad 5 = \quad 5.$$

$$9 \times 16^1 = 9 \times 16 = 144 = \underline{144.}$$

$$149.25$$

If a binary number is divided into groups of four bits, proceeding in either direction from the binary point, each group may be replaced directly by its hexadecimal equivalent. These groupings are shown in Table 8.2.

**Table 8.2** Binary/Hexadecimal Equivalents.

| Binary Grouping | Hexadecimal Digit | Binary Grouping | Hexadecimal Digit |
|---|---|---|---|
| 0000 | 0 | 1000 | 8 |
| 0001 | 1 | 1001 | 9 |
| 0010 | 2 | 1010 | A |
| 0011 | 3 | 1011 | B |
| 0100 | 4 | 1100 | C |
| 0101 | 5 | 1101 | D |
| 0110 | 6 | 1110 | E |
| 0111 | 7 | 1111 | F |

The binary number 1111001011010011 is used to illustrate the grouping of binary digits into hexadecimal digits:

| BINARY NUMBER | 1111001011010011 | | | |
|---|---|---|---|---|
| BINARY GROUPING | 1111 | 0010 | 1101 | 0011 |
| HEXADECIMAL NUMBER | F | 2 | D | 3 |

Thus the hexadecimal number F2D3 is the equivalent of the binary number 1111001011010011.

## 8.6 OCTAL NUMBERS

The octal system is based on a radix of 8 and uses the digits 0, 1, 2, 3, 4, 5, 6, and 7. The positional weights of an octal number (powers of 8) can be represented as shown in Figure 8.4.

The octal number 204.763 may be converted to its decimal equivalent by the following method:

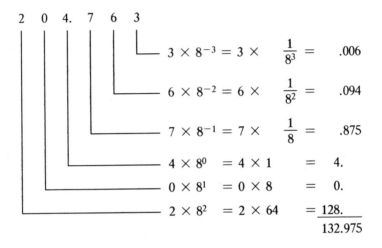

$$2 \quad 0 \quad 4. \quad 7 \quad 6 \quad 3$$

$$3 \times 8^{-3} = 3 \times \frac{1}{8^3} = .006$$

$$6 \times 8^{-2} = 6 \times \frac{1}{8^2} = .094$$

$$7 \times 8^{-1} = 7 \times \frac{1}{8} = .875$$

$$4 \times 8^0 = 4 \times 1 = 4.$$

$$0 \times 8^1 = 0 \times 8 = 0.$$

$$2 \times 8^2 = 2 \times 64 = \underline{128.}$$

$$132.975$$

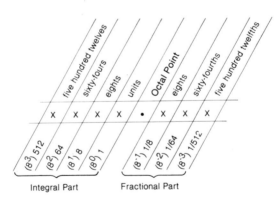

**Figure 8.4** The positional weights of an octal number (powers of 8).

**Table 8.3**  Binary/Octal Equivalents.

| Binary Grouping | Octal Digit |
|---|---|
| 000 | 0 |
| 001 | 1 |
| 010 | 2 |
| 011 | 3 |
| 100 | 4 |
| 101 | 5 |
| 110 | 6 |
| 111 | 7 |

The octal system has special characteristics that make it especially useful in many situations involving binary numbers. Since three binary digits may be grouped and represented as one octal digit, many binary operations may be represented by using octal digits. The binary groupings for the octal digits are shown in Table 8.3.

The binary number 110111101011010 is used to illustrate the grouping of binary digits into octal digits:

| BINARY NUMBER | 110111101011010 | | | | |
|---|---|---|---|---|---|
| BINARY GROUPING | 110 | 111 | 101 | 011 | 010 |
| OCTAL NUMBER | 6 | 7 | 5 | 3 | 2 |

Thus, the octal number 67532 is the equivalent of the binary number 110111101011010. It should be obvious that the writing and reading of binary numbers may be simplified by using octal notation.

## Review Exercises

1. What is the base of the following numbers?

    (a)  $76_8$  (b)  $107_{10}$  (c)  $16AF3_{16}$
    (d)  $10010010011100_2$  (e)  $10326_8$

2. Represent the first twenty-two Fibonacci numbers as binary numbers.

3. Represent the current year as a binary number.

4.  Represent your age as a hexadecimal number.

5.  Convert the binary number 10010001 to a hexadecimal number.

6.  Represent the binary number 100111101 in octal notation.

7.  Convert the binary number 10111000001 to:

    (a)  a decimal value.
    (b)  an octal value.
    (c)  a hexadecimal value.

8.  Convert octal 767 to a decimal number. Convert the octal number 4763 to equivalent binary and hexadecimal numbers.

9.  Convert the hexadecimal number 248 to:

    (a)  a binary value.
    (b)  a decimal value.
    (c)  an octal value.

10. Convert hexadecimal A49F to an equivalent decimal number.

11. Draw a flowchart and write a BASIC program to play the game of Nim with a human opponent.

12. Write a BASIC program to convert from one number base to another.

# Chapter 9

# MODULAR ARITHMETIC

**Preview**

Number theory has an algebra of its own, known as the theory of con-gruences. Ordinary algebra developed as a shorthand for arithmetic opera-tions. Similarly, congruences represent a symbolic language for divisibility, the basic concept of number theory.

After you finish this chapter, you should be able to:

1. Define congruent, modulo, and modular arithmetic.
2. Perform modular arithmetic.
3. Use the casting out nines technique to check the result of an arithmetic operation.
4. See how a computer can solve a number problem using the Chinese re-mainder theorem.

## 9.1 INTRODUCTION

Modular arithmetic, sometimes called congruence arithmetic, modulo arithmetic, or clock arithmetic, is the application of fundamental operations that involve the use of numbers of one system only. Thus, the modulo $M$ or mod $M$ system uses only the numbers $0, 1, 2, \ldots, (M - 1)$. The fundamen-tal operations are the same as those of ordinary arithmetic except that if the number is greater than $(M - 1)$ it is divided by $M$ and the remainder is used in place of the ordinary result. Generally speaking, when we write $A$ modulo $M$, we mean the remainder when $A$ is divided by $M$. More precisely, two numbers are considered congruent modulo $M$ if they leave the same remain-der. They leave the same remainder if their difference is divisible by $M$.

**Example 1.** 23 is congruent to 8 (mod 5) since $23 - 8 = 15 = 5 \cdot 3$

**Example 2.**    47 is congruent to 11 (mod 9) since $47 - 11 = 36 = 9 \cdot 4$

**Example 3.**    $81 \equiv 0$ (mod 27) since $81 - 0 = 81 = 27 \cdot 3$
The symbol $\equiv$ means "is congruent to."

## 9.2  CLOCK ARITHMETIC

Modular arithmetic may seem new, but you will find you have been applying the principles involved for some time without having any mathematical language to describe what you were doing. A familiar instance of modular arithmetic is clock arithmetic. Any two hours differing by a multiple of 12 hours are given the same numeral, so we say that the hours are counted modulo 12. In fact, any periodicity gives rise to a congruence relation. Two calendar dates separated by a multiple of 7 fall on the same day of the week. Two angles differing by a multiple of 360° have common initial and terminal sides.

Let us use the clock face shown in Figure 9.1 and illustrate how clock arithmetic is performed. In this system $2 + 11$ means start at 2, then move the hand 11 units (clockwise) to 1. In this arithmetic, $2 + 11 = 1$.

If a scientist is performing an experiment in which it is necessary to keep track of the total number of hours elapsed since the start of the experiment, he may label the hours sequentially, as 1, 2, 3, etc. When 53 hours have elapsed, it is 53 o'clock experiment time. How does he reduce experiment time to ordinary time? If zero hours experiment time corresponds to midnight, his task is easy. He simply divides by 12 and the remainder is the time of day. For example, 53 experimental time is 5 o'clock, because 12 goes into 53 with a remainder of 5; 53 is congruent to 5 modulo 12.

$$53 \equiv 5 \quad (\text{mod } 12).$$

Using a 12 hour clock, verify that each of the following is correct:

$$8 + 7 \equiv 3$$
$$5 + 12 \equiv 5$$
$$3 + 11 \equiv 2$$

The following BASIC program starts at random times and adds random numbers of hours. The random times must be numbers from 1 to 12. The random numbers of hours could have been from virtually any range, but 1 to 36 is used in this program.

```
100  REM CLOCK ARITHMETIC
110  FOR L=1 TO 8
120  LET T=INT(12*RND(1)+1)
130  LET H=INT(36*RND(1)+1)
140  LET S=T+H
150  IF S <= 12 THEN 180
160  LET S=S-12
170  GOTO 150
180  PRINT H;"HOURS FROM";T;"O'CLOCK"
190  PRINT " IT WILL BE";S;"O'CLOCK"
200  PRINT
210  NEXT L
220  END

RUN


7    HOURS FROM 5    O'CLOCK
IT WILL BE 12    O'CLOCK


27   HOURS FROM 5    O'CLOCK
IT WILL BE 8    O'CLOCK


1    HOURS FROM 2    O'CLOCK
IT WILL BE 3    O'CLOCK


4    HOURS FROM 3    O'CLOCK
IT WILL BE 7    O'CLOCK


21   HOURS FROM 2    O'CLOCK
IT WILL BE 11   O'CLOCK


5    HOURS FROM 9    O'CLOCK
IT WILL BE 2    O'CLOCK


7    HOURS FROM 8    O'CLOCK
IT WILL BE 3    O'CLOCK


29   HOURS FROM 1    O'CLOCK
IT WILL BE 6    O'CLOCK
```

START                    FINISH

**Figure 9.1**    Clock arithmetic. $2 + 11 = 1$.

The program picks eight pairs of random numbers, with T for time and H for hours, and adds them to see whether the sum is less than or equal to 12. If the sum is less than or equal to 12, the program prints the sum as the time. If the sum is greater than 12, the program determines and prints the number modulo 12. After a sum is printed, the program loops back to pick another pair of numbers, and repeats the process until 15 pairs of numbers have been picked and processed.

## 9.3   CASTING OUT NINES TECHNIQUE

An ancient technique known as casting out nines is a method used to check multiplication or addition. It is based on the excess of nines in the digits of an integer. The excess of nines is the remainder when the sum of the digits is divided by 9. For example, the excess of nines in 9357 is 6, since $9 + 3 + 5 + 7 = 24$, and the remainder when 24 is divided by 9 is 6. The sum of several numbers can be checked as follows:

$$
\begin{array}{rl}
\$\ 14.32 & \quad 1 + 4 + 3 + 2 \equiv 1 \quad (\mathrm{mod}\ 9) \\
1.48 & \quad 1 + 4 + 8 \equiv 4 \quad (\mathrm{mod}\ 9) \\
76.25 & \quad 7 + 6 + 2 + 5 \equiv 2 \quad (\mathrm{mod}\ 9) \\
75.65 & \quad 7 + 5 + 6 + 5 \equiv 5 \quad (\mathrm{mod}\ 9) \\
\underline{4.13} & \quad 4 + 1 + 3 \equiv \underline{8} \quad (\mathrm{mod}\ 9) \\
\$171.83 & \quad 1 + 7 + 1 + 8 + 3 \equiv 2 \quad (\mathrm{mod}\ 9)
\end{array}
$$

This method can be applied to any arithmetical expression involving any combination of additions and multiplications. The casting out nines technique is based on congruence properties modulo 9.

This same idea can be used to test a large number for divisibility by 9. Is 5438777 evenly divisible by 9?

$$5438777 = 5 + 4 + 3 + 8 + 7 + 7 + 7$$

$$\equiv 5 \not\equiv 0 \quad (\mathrm{mod}\ 9),$$

therefore, 5438777 is not evenly divisible by nine.

The casting out nines technique can also be used to form pretended mind reading tricks you can use to entertain your friends.

Begin by asking your friend to write down any number, containing as many digits as he or she wishes, such as 6347.

$$
\begin{array}{r}
6347 \\
\underline{7436} \\
1089
\end{array}
$$

Then tell him or her to write the same number backward (7436). Then tell him to subtract the smaller from the larger (yielding 1089). Next tell him or her to cross out any one digit of his or her choice in this number except that it must not be a zero. Then ask him or her to read you the remaining digits, and you will tell him or her the digit he or she crossed out. (In this example, if he or she crosses out the 8, he or she then reads you the digits "109." You immediately tell him or her: "You crossed out an 8").

The trick is performed as follows: when you hear the remaining digits, you mentally add them together $(1 + 0 + 9 = 10)$. If the result contains more than one digit, add them together $(1 + 0 = 1)$. Continue until you have only a single digit (1 in this case), then subtract it from 9, $(9 - 1 = 8)$. The answer is the digit crossed out. An exception occurs when the sum of the digits is 9. In this case the crossed out digit is the sum itself, 9.

There are many interesting variations of this demonstration. For example, you can ask your friend to write any number containing any number of digits, then to write the sum of the digits below and subtract, then to cross out any one of the nonzero digits in the result, as shown:

$$\begin{array}{r} 763218 \\ -27 \\ \hline 763191 \end{array}$$

Another variation is to give your friend these instructions:

1. Write down any positive three-digit integer in which the first digit differs from the third by at least two, 265 for example.
2. Now write the same number backwards, below the original number (562).
3. Subtract the smaller of these numbers from the larger (297).
4. Reverse the result (792), and add it to the number you had before you reversed it $(792 + 297 = 1089)$.

The example shows the steps for 265.

$$\begin{array}{r} 265 \\ 562 \\ \hline 297 \\ 792 \\ \hline 1089 \end{array}$$

At this point you can ask your friend to be quiet while you meditate to discover his final result by ESP. After a minute or so, you announce the answer: 1089. This will always be correct if he has carried out the steps correctly, no matter what number he started with.

## 9.4  CHINESE REMAINDER THEOREM

The Chinese remainder theorem states the following result: given a set of congruences of the form

$$X \text{ is congruent to } A \quad (\text{mod } M)$$

where each pair of $M$'s is relatively prime (that is, it has no factor greater than 1 in common), if we let $P$ be the product of the $M$'s, we then have a unique value $B$, $0 \leq B < P$ whereby every $X$ having all the given properties is congruent to $B$ modulo $P$.

This theorem is used in an old mind reading trick when the huckster asks someone in the audience to think of a number between one and 30. Then his spiel goes: "Don't tell me the number, but divide the number by two and give me the remainder, next divide the number by three and give me the remainder, and last, divide the number by five, and give me the remainder." When the huckster has the three remainders he is able to calculate the original number.

Take the number 12, for example:

$$12/2 = 6 \text{ Remainder } 0$$

$$12/3 = 4 \text{ Remainder } 0$$

$$12/5 = 2 \text{ Remainder } 2$$

Given these three remainders, it is easy to determine the original number since there is only one number less than 30 with these three remainders. Try it for yourself. Note that the three divisors are relatively prime (no common factors within the range $2 \times 3 \times 5 = 30$).

The following program uses the Chinese remainder theorem with the computer playing the part of the huckster. In the program 5, 7, 9 are used for the divisors (Range $= 5 \times 7 \times 9 = 315$).

```
100  REM CHINESE REMAINDER THEOREM
101  PRINT "I WANT YOU TO THINK OF A NUMBER"
102  PRINT "LESS THAN 316. WRITE THIS NUMBER"
103  PRINT "DOWN AND DIVIDE BY 5. NOW GIVE ME"
104  PRINT "THE REMAINDER LEFT OVER";
110  INPUT R5
111  PRINT
120  PRINT "NOW DIVIDE YOUR ORIGINAL NUMBER BY"
121  PRINT "7 AND GIVE ME THIS REMAINDER";
130  INPUT R7
131  PRINT
140  PRINT "NOW DIVIDE YOUR ORIGINAL NUMBER BY"
```

```
141   PRINT "9 AND GIVE ME THIS REMAINDER";
150   INPUT R9
151   PRINT
160   REM CALCULATE NUMBER
170   LET A=126*R5+225*R7+280*R9
180   LET X=A-INT(A/315)*315
190   PRINT
200   PRINT "I AM HAPPY TO TELL YOU THAT YOUR"
201   PRINT "NUMBER CHOSEN WAS";X
210   END

RUN


I WANT YOU TO THINK OF A NUMBER
LESS THAN 316. WRITE THIS NUMBER
DOWN AND DIVIDE BY 5. NOW GIVE ME
THE REMAINDER LEFT OVER?0

NOW DIVIDE YOUR ORIGINAL NUMBER BY
7 AND GIVE ME THIS REMAINDER?1

NOW DIVIDE YOUR ORIGINAL NUMBER BY
9 AND GIVE ME THIS REMAINDER?8


I AM HAPPY TO TELL YOU THAT YOUR
NUMBER CHOSEN WAS 260
```

In the output example, the number 100 was chosen. The algorithm for calculating the number is:

$$X = (126 \times R5 + 225 \times R7 + 280 \times R9) \mod 315$$

$$X = (126 \times 0 + 225 \times 2 + 280 \times 1) \mod 315$$

$$X = (730) \mod 315$$

$$X = 100$$

X is the number. R5 is the remainder when X is divided by 5, R7 is the remainder when X is divided by 7, and R9 is the remainder when X is divided by 9. Mod 315 indicates remainder arithmetic, where the parenthesized quantity is divided by 315 but the result is the remainder, not the quotient.

The algorithm in the program is not exactly of this form since BASIC does not have a modulus function.* However, BASIC does have an INTEGER function, which is used in the program (line number 180) to do the remainder arithmetic.

## 9.5 MONKEYS AND COCONUTS

You have probably already encountered the monkeys and coconuts problem that has tantalized many a problem solver.

On an island in the Pacific Ocean, five shipwrecked sailors plan to divide a pile of coconuts among themselves in the morning. During the night one of them wakes up and decides to take his share. After throwing a coconut to a monkey to make the division come out evenly, he takes one fifth of the pile and goes back to sleep. The other four sailors do likewise, one after the other, each throwing a coconut to the monkey and taking one fifth of the remaining pile. In the morning the five sailors throw a coconut to the monkey and divide the remaining coconuts into five equal piles. What is the smallest number of coconuts that could have been in the original pile?

The solution is as follows. If the original number of coconuts is $N$ and the number each sailor receives in the final division is $A$, then

$$\frac{1}{5}\left(\frac{4}{5}\left(\frac{4}{5}\left(\frac{4}{5}\left(\frac{4}{5}(N-1)-1\right)-1\right)-1\right)-1\right)-1\right) = A.$$

Removing the parentheses and regrouping yields

$$\left(\frac{4}{5}\right)^5 N - \left[1 + \frac{4}{5} + \left(\frac{4}{5}\right)^2 + \left(\frac{4}{5}\right)^3 + \left(\frac{4}{5}\right)^4 + \left(\frac{4}{5}\right)^5\right] = 5A.$$

Summing the series and clearing of fractions results in

$$4^5(N+4) = 5^6(A+1).$$

---

*Some versions of the BASIC language include a library function to perform modulus arithmetic.

The prime factors of $5^6$, which are $5 \times 5 \times 5 \times 5 \times 5 \times 5$, must be contained in the lefthand side of the previous equation. Since none of them is in $4^5$, $5^6$ must divide $(N + 4)$:

$$(N + 4) \text{ is divisible by } 5^6 = 15625.$$

or,

$$N \equiv -4 \quad (\text{mod } 15625).$$

But we can't start with $-4$ coconuts, and the very next integer in this residue class modulo 15625 is $-4 + 15625 = 15621$. That's a big pile of coconuts!

### Review Exercises

1.  Solve each of the following for $x$ in arithmetic modulo 5.

    (a)   $1 - 3 = x$
    (b)   $2 - 4 = x$
    (c)   $3 - x = 4$
    (d)   $2 - x = 4$

2.  Write a BASIC program that computes and prints a multiplication table modulo $M$. That is, it should contain the product of each pair of numbers from among the numbers 1, 2, 3, ..., $M - 1$. Print the tables for $M = 4, 6, 8, 10$, and 12.

3.  Write a program to type an addition table and a multiplication table modulo $N$, where $N$ is any number less than 20.

4.  Consider arithmetic on a four minute clock with the elements (0, 1, 2, 3), as in the figure. Create addition tables for arithmetic modulo 4.

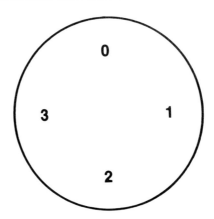

5.  Solve the equation $T + 6 = 2$ for $T$ where $T$ may be replaced by any one of the numerals on a 12 hour clock.

6.  What is $3 - 7$ on the 12 hour clock? Note that $3 - 7 = T$ is equivalent to $T + 7 = 3$.

7.  Make a complete table of addition facts on the 12 hour clock.

8.  Solve each equation where $T$ may be replaced by any one of the numerals on a 12 hour clock:

    (a)  $T - 3 = 11$
    (b)  $3 + T = 2$
    (c)  $T + 7 = 5$

9.  Draw a flowchart that could have been used to write the BASIC program given in Section 9.2.

10. As shown in Section 9.3, a number is divisible by 9 only if the sum of its digits is divisible by 9. Determine if 234 648 is divisible by 9.

11. Write a BASIC program to determine if the number 39 827 437 is divisible by 9. The program should use the casting out nines technique.

12. Check the addition of the following numbers using the casting out nines technique.

$$
\begin{array}{r}
14\ 745 \\
23\ 610 \\
10\ 100 \\
21\ 007 \\
6\ 143 \\
\underline{12\ 841} \\
88\ 446
\end{array}
$$

13. Draw a flowchart that could have been used to write the BASIC program given in Section 9.4.

14. Draw a flowchart and write a BASIC program to print out a calendar in a familiar form in response to an integer representing the year (e.g., 1976). To calculate the day of a week, use the following formula:

$$F = \{(2.6M - 0.2) + K + D + (D/4) + (C/4) - 2C\} \bmod 7$$

where braces represent the integer part, and mod 7 implies the remainder after division by 7. Variables are defined as follows:

$F$ = day of week (Sunday = 0, Monday = 1, ...)

$K$ = day of month (1, 2, 3, ..., 31)

$C$ = century (18, 19, 20, ...)

$D$ = year in century (1, 2, 3, ..., 99)

$M$ = month number with January and February taken as months 11 and 12 of the preceding year (March = 1, April = 2, ...). Thus January 7, 1975, has $K = 7$, $C = 19$, $D = 74$, $M = 11$. Note: This formula applies only to years after the calendar change in 1752.

# Chapter 10

# NUMBER THEORY FOR FUN

**Preview**

If you have ever solved a mathematical puzzle, played any game in which numbers were used, constructed a magic square, or learned a number trick, you have engaged in a type of mathematical recreation. In this chapter are many interesting and fun problems taken from the field of number theory.

After you complete this chapter, you should be able to do the following:

1. Devise number tricks that will fool your friends.
2. Define Diophantine problems, Pythagorean triplets, and Pascal's triangle.
3. See how a computer can be used to generate Pythagorean triplets and Pascal's triangle.
4. Understand two famous unsolved problems in the field of number theory: Fermat's last theorem, and Goldbach's conjecture.

## 10.1  MIND READING TRICKS

In Section 1.2 we presented a simple mathematical number trick. In Section 9.3 we saw how the casting out nines technique could be used to perform pretended mind reading tricks. In this section are a few more tricks, all based upon fairly simple mathematical operations. You may wish to let the computer become the mind reader by programming one or more of the examples.

$* * *$ Trick 1 $* * *$

The mind reader (computer) asks a person in his audience to think of a number, multiply it by 5, add 6, multiply by 4, add 9, multiply by 5, and state the result.

The person chooses the number 12, calculates successively 60, 66, 264, 273, 1365, and announces the last number.

The mind reader (computer) subtracts 165 from the result, gets 1200, knocks off the two zeros, and informs the person that 12 was his original number.

The trick is easily seen if put in mathematical symbols. If the number chosen is $a$, then the successive operations yield

$$5a$$
$$5a + 6$$
$$20a + 24$$
$$20a + 33 \text{ and}$$
$$100a + 165$$

When the mind reader (computer) is told this number, it is evident that he can determine $a$ if he subtracts 165 and then divides by 100.

### * * * Trick 2 * * *

If the mind reader (computer) desires to tell a person the result without asking any questions, he must arrange the various operations so that the original number drops out. Here is an example in which three unknown numbers are introduced and eliminated.

The mind reader (computer) says: Think of a number. Add 10. Multiply by 2. Add the amount of change in your pocket. Multiply by 4. Add 20. Add 4 times your age in years. Divide by 2. Subtract twice the amount of change in your pocket. Subtract 10. Divide by 2. Subtract your age in years. Divide by 2. Subtract your original number.

The person, who chooses the number 7, has 30 cents in his pocket, and is 20 years old, thinks: 7, 17, 34, 64, 256, 276, 356, 178, 118, 108, 54, 34, 17, 10.

The mind reader (computer) says your result is 10, is it not? the person replies, "Right!"

In this case, if we denote the person's original number by $a$, the amount of change in his pocket by $b$, and his age in years by $c$, the successive operations give

$$a$$
$$a + 10$$
$$2a + 20$$
$$2a + 20 + b$$
$$8a + 80 + 4b$$

$$8a + 100 + 4b + 4c$$
$$4a + 50 + 2b + 2c$$
$$4a + 50 + 2c$$
$$4a + 40 + 2c$$
$$2a + 20 + c$$
$$2a + 20$$
$$a + 10$$
$$10$$

Problems of this type can be set up in many ways.

### * * * Trick 3 * * *

Many tricks of the kind we are discussing are based upon the principle of positional notation. Consider the following.

The mind reader (computer) says: throw three dice and note the three numbers which appear. Operate on these numbers as follows: Multiply the number on the first die by 2, add 5, multiply by 5, add the number on the second die, multiply by 10, add the number on the third die, and state the result.

The person throws a 2, a 3, and a 4, and thinks: 4, 9, 45, 48, 480, 484. He gives the answer 484.

The mind reader (computer) subtracts 250 and gets 234. He then states that the numbers thrown were 2, 3, and 4.

### * * * Trick 4 * * *

The mind reader (computer) says: choose any prime number greater than 3, square it, add 17, divide by 12, and remember the remainder.

The person thinks: 11, 121, 138, 116/12, 6.

The mind reader (computer) states that the remainder is 6.

Here use is made of the fact that any prime number greater than 3 is of the form $6n \pm 1$, where $n$ is a whole number. The symbol $\pm$ means plus or minus, its square is then of the form $36n^2 \pm 12n + 1$.

This number, when divided by 12, leaves a remainder of 1. Now the mind reader (computer) had the person add 17, which, divided by 12, leaves a remainder of 5. The final remainder must thus be $1 + 5$, or 6.

The mind reader (computer) can vary this trick by using a number other than 17. The number when divided by 12, will have a remainder equal to $k$. Then the final remainder will always be $1 + k$.

## 10.2  MATHEMAGIC SQUARE

Closely akin to magic squares are square arrays of numbers such as the following. We begin by forming a 3 by 3 square array and placing any six integers in the surrounding head spaces as shown. The numbers 6, 4, 3, 1, 2, and 7 are chosen arbitrarily.

| + | 6 | 4 | 3 |
|---|---|---|---|
| 1 | | | |
| 2 | | | |
| 7 | | | |

Next find the sum of each pair of numbers as in a regular addition table.

| + | 6 | 4 | 3 |
|---|---|---|---|
| 1 | 7 | 5 | 4 |
| 2 | 8 | 6 | 5 |
| 7 | 13 | 11 | 10 |

Now let us perform the trick. Have your friend circle one of the nine numbers in the box, say 5, and then cross out all the other numbers in the same row and column as 5.

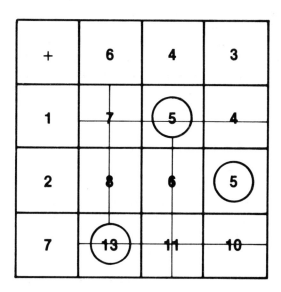

Next circle one of the remaining numbers, say 13, and repeat the process. Circle the only remaining number, the 5. The sum of the circled numbers is 5 + 13 + 5 = 23.

The sum of the three circled numbers will always be 23. Furthermore, the sum of the six numbers outside the square is 23. Try to explain why this trick works and build a table with 16 entries.

## 10.3  DIOPHANTINE PROBLEMS

If you are interested in puzzles, you are probably familiar with problems of the following type. A student's transcript shows $T$ 3-hour courses and $F$ 5-hour courses, for a total of 64 hours. Find $T$ and $F$. From the description of the problem we get the equation

$$3T + 5F = 64.$$

This one equation, in two unknowns, constitutes the entire translation into algebra of the conditions. However, there is a difference between this problem and those considered in previous chapters. Here we want only integral solutions. To solve this problem experimentally, note that $F$ must be less than 13, no matter what $T$ is, since $5 \times 13 = 65$, and 65 is greater than 64. If $F = 12$ we get $3T = 4$, which is not a solution. Continuing to decrease $F$, we find four possible solution pairs, corresponding to the four points on the line $3T + 5F = 64$ in Figure 10.1.

Equations like the previous one that have to be solved for integral values of the unknowns are called Diophantine equations, after the Greek mathematician Diophantus. Frequently such an equation has no solution. For example, $6X + 9Y = 16$ has no solution in integers, since the lefthand side is divisible by 3 and the righthand side is not.

### 10.4    FOUR COLOR PROBLEM

When coloring a geographical map, it is customary to give different colors or different shades to any two countries that have a portion of their boundary in common. It has been found empirically that any map, no matter how many countries it contains or how they are situated, can be so colored by using only four different colors. It is easy to see that no smaller number of colors will suffice for all cases (see Figure 10.2).

The four color problem is one of the most celebrated challenges in mathematics. It is of great intellectual interest and has intrigued many people from all paths of life. Its solution has little or nothing whatsoever to do

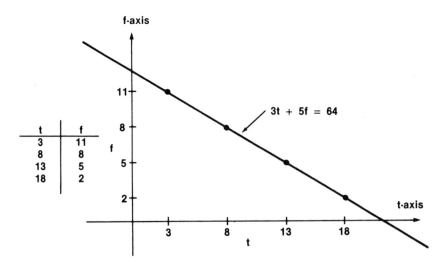

**Figure 10.1**   Experimental solution of $3t + 5f = 64$.

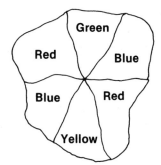

**Three colors are sufficient for this map**

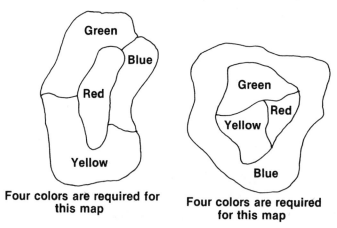

**Four colors are required for this map**

**Four colors are required for this map**

**Figure 10.2**   Coloring a map.

with making maps. A map maker is and always will be able to print maps using as many different colors as he needs.

No one has ever been able to produce a map that would require more than four colors, and until 1976 no one had been able to prove that four colors are sufficient for all maps. The computer-aided proof of the century-old conjecture was completed in 1976 and published in 1977, by Professors V. Appel and W. Haken at the University of Illinois. This proof cannot in fact be checked by hand calculations. In more than a century of research no simple elegant proof of this problem has ever been demonstrated. But the search for such a proof has stimulated development of new branches of mathematical science in the fields of combinatorial mathematics and topology.

Two amazing revelations appeared in the 1977 papers of Appel and Haken. The first was that these two mathematicians had finally solved the problem that had frustrated all attempts for one and a quarter centuries. The second was even more significant in its implications for the world of mathematics: the successful proof was dependent upon the use of a high-speed computer and could not have been accomplished without it. The computer-assisted proof of the four color conjecture demanded about 10 billion logical decisions and required more than 1200 hours of work analyzing thousands of configurations by computer.

## 10.5   THE PYTHAGOREAN PROBLEM

An early number theory is the Pythagorean problem. As you know, in a right-angled triangle the lengths of the sides satisfy the Pythagorean relation

$$X^2 + Y^2 = Z^2,$$

where $Z$ is the length of the hypotenuse, the side opposite the right angle.

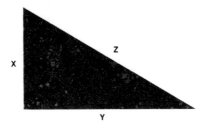

This equation makes it possible to compute the length of one side of a right triangle when you know the other two.

Sometimes all the side lengths $X$, $Y$, $Z$ (reference diagram) are integers. The simplest case is $X = 3$, $Y = 4$, and $Z = 5$. There are many other cases of integral solutions of the Pythagorean equation:

| | | |
|---|---|---|
| $X = 5$ | $Y = 12$ | $Z = 13$ |
| $X = 7$ | $Y = 24$ | $Z = 25$ |
| $X = 8$ | $Y = 15$ | $Z = 17$ |
| $X = 9$ | $Y = 40$ | $Z = 41$ |
| $X = 11$ | $Y = 60$ | $Z = 61$ |
| $X = 12$ | $Y = 35$ | $Z = 37$ |
| $X = 13$ | $Y = 84$ | $Z = 85$ |

$$X = 20 \qquad Y = 21 \qquad Z = 29$$
$$X = 28 \qquad Y = 45 \qquad Z = 53$$
$$X = 33 \qquad Y = 56 \qquad Z = 65$$

These sets of three integers, $X$, $Y$, and $Z$, are called Pythagorean triplets. An easy way to produce several sets of Pythagorean triplets is to use the equations

$$X = 2UV$$

$$Y = U^2 - V^2$$

$$Z = V^2 + U^2$$

where $U$ and $V$ are relatively prime (the only common factor is 1), where $V < U$, and where one of the numbers $U$ and $V$ is even and the other is odd.

The following BASIC program uses the previous equations to generate several sets of triplets. The program varies $U$ from 1 to 6 and $V$ from 2 to 7. The program prints multiple triplets as well as original triplets. For example, the triplets (10, 24, 26) and (20, 48, 52) are multiples of (5, 12, 13).

```
100   REM PYTHAGOREAN TRIPLES
110   PRINT "PYTHAGOREAN TRIPLES"
120   PRINT "TRIPLETS 16N U SN V"
130   FOR U=1 TO 6
140   FOR V=U+1 TO 7
150   REM GENERATE TRIPLES
160   LET A[1]=2*U*V
170   LET A[2]=ABS(U↑2-V↑2)
180   LET A[3]=V↑2+U↑2
190   REM PRINT TRIPLETS, U, AND V
220   PRINT A[1];A[2];A[3],U;V
230   NEXT V
240   NEXT U
250   END

RUN

PYTHAGOREAN TRIPLES
      TRIPLETS                      U     V
   4    3    5                      1     2
   6    8    10                     1     3
```

| 8  | 15 | 17 |  | 1 | 4 |
|----|----|----|--|---|---|
| 10 | 24 | 26 |  | 1 | 5 |
| 12 | 35 | 37 |  | 1 | 6 |
| 14 | 48 | 50 |  | 1 | 7 |
| 12 | 5  | 13 |  | 2 | 3 |
| 16 | 12 | 20 |  | 2 | 4 |
| 20 | 21 | 29 |  | 2 | 5 |
| 24 | 32 | 40 |  | 2 | 6 |
| 28 | 45 | 53 |  | 2 | 7 |
| 24 | 7  | 25 |  | 3 | 4 |
| 30 | 16 | 34 |  | 3 | 5 |
| 36 | 27 | 45 |  | 3 | 6 |
| 42 | 40 | 58 |  | 3 | 7 |
| 40 | 9  | 41 |  | 4 | 5 |
| 48 | 20 | 52 |  | 4 | 6 |
| 56 | 33 | 65 |  | 4 | 7 |
| 60 | 11 | 61 |  | 5 | 6 |
| 70 | 24 | 74 |  | 5 | 7 |
| 84 | 13 | 85 |  | 6 | 7 |

Look at the list of triplets produced by the program. Do you find any interesting relationships? Did you know that every Pythagorean triplet has at least one element that is divisible by either 3, 4, or 5? For example, (12, 35, 37) has 12, which is divisible by both 3 and 4, and 35, which is divisible by 5. Notice that the product of all three numbers is always a multiple of 60.

## 10.6 PASCAL'S TRIANGLE

Determining the chance that something will happen is like looking into the future. It is done by using common sense and a knowledge of what happened in the past. To see how it works in a simple case, let us try to foresee what happens when you toss a coin. The coin has two faces, head and tail. Common sense and experience join to tell us that, out of a large number of tosses, about half will come out heads, and the rest will be tails. Saying it another way: on the average, one out of two tosses will come out heads. So we say the chance of getting a head is ½.

If we toss two coins, there are three possible results. We may get two heads, or two tails, or one head and one tail. The chance of getting two heads is one out of four, or ¼. The chance of getting two tails is also ¼. The chance of getting one head and one tail is two out of four, or ½. What is the chance of getting two heads and a tail when you toss three coins? If you toss four coins, what is the chance of getting all heads or all tails?

**PASCAL**

There is a short method for finding the answers to these questions in the arrangement of numbers known as Pascal's triangle. Pascal, a French philosopher and mathematician of the seventeenth century, was for a time interested in games of chance. This interest led him to discover certain important rules about the probabilities of getting heads or tails on the toss of a coin. His findings are described in the following triangular formation of numbers. The formation is easy to construct and shows the chance of getting heads or tails, or any combination of them, on a given number of tosses of a coin.

```
                          1
                       1     1
                    1     2     1
                 1     3     3     1
              1     4     6     4     1
           1     5    10    10     5     1
        1     6    15    20    15     6     1
     1     7    21    35    35    21     7     1
  1     8    28    56    70    56    28     8     1
1     9    36    84   126   126    84    36     9     1
1   10    45   120   210   252   210   120    45    10    1
1   11    55   165   330   462   462   330   165    55    11    1
```

If you toss one coin, the chance of getting heads is 1 out of 2, or ½. If you toss two coins, your chance of getting 2 heads is 1 out of 4; of getting 1 head and 1 tail, 2 out of 4 or ½; of getting 2 tails, 1 out of 4. If you toss three coins, your chances are: all heads, 1 out of 8; 2 heads and 1 tail, 3 out of 8; 2 tails and 1 head, 3 out of 8; all tails, 1 out of 8.

If four coins are tossed, there is 1 chance in 16 of getting all heads or all tails; 4 out of 16 of getting 3 heads and 1 tail, or 3 tails and 1 head; and 6 out of 16 of getting 2 heads and 2 tails. In five tosses, chances are: 1 out of 32 for all heads or all tails; 5 out of 32 for 4 heads and 1 tail, or 4 tails and 1 head; and 10 out of 32 for 3 heads and 2 tails or 3 tails and 2 heads.

There are many different ways to generate the triangle: combinatorial methods, trigonometric methods, or the binomial theorem method. A very easy way of constructing the triangle exists. Each number within the triangle is found by adding the two numbers above it at the left and right. The coefficients in the expansion of $(A + B)^N$ as $N$ is assigned the values 0, 1, 2, ... successively can be used to form the triangular array. The number of combinations of $N$ things taken $R$ at a time (a basic problem in probability theory) can also be used to produce the coefficients of this expansion.

The following BASIC program produces the numbers of Pascal's triangle.

```
10   PRINT "PASCAL'S TRIANGLE"
20   FOR N=0 TO 7
30   FOR R=0 TO N
40   LET C=1
50   FOR X=N TO N-R+1 STEP -1
60   LET C=C*X/(N-X+1)
70   NEXT X
80   PRINT C;
90   NEXT R
100   PRINT
110   NEXT N
120   END

RUN


PASCAL'S TRIANGLE
1
1    1
1    2    1
1    3    3    1
1    4    6    4    1
1    5    10   10   5    1
1    6    15   20   15   6    1
1    7    21   35   35   21   7    1
```

## 10.7   UNSOLVED PROBLEMS

Probably no branch of mathematics other than number theory is so replete with unsolved problems that are so simple as to be readily understood by someone without advanced training. Two of the more famous unsolved problems are described in this section. Other unsolved problems are described in the number theory books listed in the bibliography.

**Fermat's Last Theorem.**   None of the famous unsolved problems outrank Fermat's last theorem, the proposition that the equation

$$x^n + y^n = z^n$$

has integral solutions in $x$, $y$, $z$ for $n = 1$, 2, and for no number greater than 2. The great mathematician Fermat claimed to have found a proof of the impossibility of integral solutions for this equation, a proof which, he said, "the margin of my book is too small to contain." Repeated assaults on this problem by eminent mathematicians have resulted in proofs that Fermat's claim of impossibility is correct for many special values of $n$. To this day, however, no proof for all values of $n$ has been found. But the search has not been in vain, for other far-reaching discoveries have been made as by-products of the research. Perhaps you would like to write a BASIC program to produce several solutions to this equation. It has many solutions—$3^2 + 4^2 = 5^2$, $5^2 + 12^2 = 13^2$, or $8^2 + 15^2 = 17^2$. What, then, of the equations $x^3 + y^3 = z^3$, $x^4 + y^4 = z^4$, and so forth?

**Goldbach's Conjecture.**   Christian Goldbach (1690–1764), a Russian mathematician, made a shrewd guess—in mathematics it is called a conjecture—that every even number is the sum of two prime numbers. For instance, $12 = 5 + 7$ and $18 = 5 + 13$.

Goldbach communicated his guess to his illustrious friend Leonhard Buler (1707–1783), a Swiss mathematician, who was quite impressed. The surmise seemed to him to be a true proposition, but his sustained efforts to prove it and the efforts of his followers up to the present time were in vain.

In the late 1930s, I. M. Vinogradov proved that any odd number is the sum of three prime numbers. Thus $17 = 3 + 7 + 7$, $19 = 3 + 5 + 11$, $21 = 3 + 7 + 11$, and $35 = 5 + 7 + 23$. The complete problem is still unsolved. Perhaps you would like to write a BASIC program to show that several hundred even numbers greater than 2 are the sum of two primes.

## 10.8   π  IN THE COMPUTER AGE

People have been calculating the value of π for years. Even the digit hunters of Babylon in 2000 B.C.E. knew the fundamental constant was about 3⅛. But it wasn't until the dawn of the computer age that π could be calculated to hundreds of thousands of decimal places. By 1967, the value of π was known to 500 000 decimal places. The computer needed about 28 hours to churn out these half million digits.

The first computer calculation of π was apparently made on ENIAC in 1949. It calculated π to 2037 places in 70 hours. In 1954, the NORC (Naval Ordnance Research Calculator) at Dahlgren, Virginia computed π to 3089 places in only 13 minutes. Four years later, a Pegasus computer in London, England computed π to 10 021 decimal places in 33 hours. In 1958, an IBM 704 computer was programmed to produce π to 10 000 decimal places in 1 hour and 40 minutes. In the 1960s, several computers computed π to more than 100 000 decimal places. In 1967, a Control Data 6600 computer was programmed by M. Gillond and Michele Dechampt to produce π to 500 000 decimal places. The computer running time was slightly more than 28 hours. The half-million digit values of π were published in reports of the Commissariat de l'Energie Atomique in Paris, France. It is believed that Gillond extended his computation to one million places, but no published version of the result can be found.

The poster shown in Figure 10.3 is a reproduction of the first 8128 decimal places computed by Daniel Shanks and John Wrench in 1961. (They computed π to 100 266 places in 8 hours and 43 minutes using an IBM 7090 computer system.)

## 10.9   HARD PROBLEMS

Computers have expanded mathematicians' horizons, allowing them to make calculations never before dreamed of. But, at the same time, they have made mathematicians recognize the limits of their ability to solve certain types of problems. Within the past decade, mathematicians and computer scientists grouped together hundreds of related problems. These, in principle, can be solved primarily by adding and multiplying. However, even the best methods of solving these problems can require billions upon billions of calculations—enough to keep the computers busy for years, even centuries. The scientists are now learning to live with this impediment. And perhaps they can even exploit it to create a new kind of seemingly unbreakable secret code.

**Figure 10.3** A chart showing $\pi$ to 8182 places. The above two-color chart is available from Creative Publications, Inc. P.O. Box 10328, Palo Alto, CA 94303.

These simple but possibly unsolvable problems are not new. Many have been around for decades. They crop up in many practical situations. But until 1971 mathematicians and computer scientists did not realize that the problems were related. Then, Stephen Cook of the University of Toronto made a discovery. He found that several of these problems were equivalent. This means that if anyone could find a shortcut to solving one of them, the shortcut could be adapted to solve the others.

Previously mathematicians had been looking at each problem separately, hoping somehow to find a way to solve it in a feasible length of computer time. But Cook's discovery systematized the study of these problems. Shortly afterwards, Richard Karp of the University of California at Berkeley greatly extended the list of equivalent problems. Then, in the scientific community, a scramble to find which problems were equivalent to these hard ones began. So far, hundreds have been added and more are under consideration.

Hard problems are technically called NP-complete (NP means nondeterministic polynomial) by mathematicians and computer scientists. What sort of problems are classified hard?

One example is the traveling salesman problem: a salesman wants to plan a tour of a number of cities so that he visits each city only once and wants to find the shortest possible route. This problem turns up in numerous guises in practical situations. The telephone company must solve a traveling salesman problem when it plans collections from pay telephone booths. The telephone company divides each city into zones. Each zone contains several hundred coin boxes. The company supervisors must decide the best order to collect coins from the telephones in each zone.

Another hard problem is the bin packing problem: suppose there are a given number of identical bins and a group of odd-shaped packages. What is the minimum number of bins necessary so that each package is in a bin, and none of the bins overflows? It is a bin packing problem to decide how to schedule television commercials to fit in one minute time slots. It is also a bin packing problem to find out how to cut up the minimum number of standard length boards to produce pieces of particular lengths.

The only general method of solving these hard problems is to try all possible solutions until you find the best one. For small problems, this isn't too hard. If a salesman had to visit only four cities, he could plan his tour himself with just a pencil and paper by considering all 24 possibilities.

But if the salesman had to visit ten cities, it would be considerably harder to try out all routes, because there would be more than 3 500 000 of them. The task is, however, well within the capabilities of a computer. But if a salesman had to visit 18 cities, and he had a computer that could test one million routes per second, it would take the computer about 4000 years to try all possible routes. If the salesman had to visit 60 cities, it would take a computer billions of centuries to try out all the routes (Figure 10.4).

Why does it take so long to try out all possible solutions to a hard problem? The reason is that the number of possible solutions increases explosively as the size of the problem grows. In a bin packing problem, for example, this rate of increase is found by multiplying a certain fixed number by itself each time you add another object to be packed. The fixed number equals the number of bins. The number of multiplication steps required varies with the number of objects. Thus, if you have two bins and two objects, the number of possible solutions is four (2 × 2); with three objects, there are eight possible solutions (2 × 2 × 2). Each object you add multiplies the number of possible solutions.

**Figure 10.4** Mathematicians and computer scientists are attempting to produce methods that will allow hard problems to be solved. Today, a computer might have to work longer than the universe has existed to solve some hard problems using the exponential time method.

In a bin packing problem of this kind, computer scientists have calculated that with ten objects, the possible solutions could be tried in 1/1000 of a second. With 20 objects, trying all the solutions would take about one second, with 30 objects, about 18 minutes would be needed, and with 100 objects to pack, trying all the possible solutions would take more than 400 000 billion centuries. (By contrast, the entire universe is only about 10 billion years old.)

Because the hard problems are of practical importance, many computer scientists and mathematicians are spending considerable amounts of time trying to develop methods that will solve some of them. Obviously, exponential time methods are not practical for large problems. What mathematicians would like to find is a polynomial time method that would require much less computer time. Computer scientists calculate that a problem that would take billions of centuries to solve by an exponential method might be solved in seconds if a polynomial method were found.

An exciting spin-off from the discovery of hard problems is the idea of using them as the basis of a new kind of secret code. Such a code was first suggested a few years ago by Whitfield Diffie, Martin Hellman, and Ralph Merkle of Stanford University. The reason the code may be unbreakable is that an eavesdropper trying to decipher a message would have to solve a hard problem to do so.

### Review Exercises

1. Construct a variation of one of the mind reading puzzles given in Section 10.1.

2. Write a BASIC program to build a Mathemagic Square with 16 entries and play this game with a human player.

3. Find a solution in integers of $5x + 7y = 29$.

4. Find a solution in integers of $33x + 14y = 173$.

5. Modify the BASIC program given in Section 10.5 to include all values $V \le 10$.

6. Draw a flowchart that could have been used to write the BASIC program given in Section 10.5.

7. Write a program to find and print all Pythagorean triangles whose hypotenuse is $\le 200$.

8. Write a BASIC program to find all Pythagorean triangles with one side equal to 12.

9. Draw a flowchart that could have been used to write the program given in Section 10.6.

10. Draw a flowchart and write a BASIC program to compute and print the numbers in Pascal's triangle. The program can use the fact that each number, other than 1, is the sum of the two closest numbers in the previous line.

11. Write a BASIC program to produce several solutions to Fermat's last theorem.

12. Write a BASIC program to show that the even numbers between 2 and 300 are the sum of two primes.

13. What is meant by a "hard problem?"

# GLOSSARY OF COMPUTER TERMS

**Acronym** An identifying word or expression from initials or segments of a name, term, or phrase, e.g., BASIC from *B*eginners *A*ll-purpose *S*ymbolic *I*nstruction *C*ode.

**Algorithm** A list of instructions specifying a sequence of operations that will give the answer to any problem of a given type.

**Alphanumeric** Consisting of alphabetical and numerical characters combined.

**Analysis** The investigation of a problem by a consistent method.

**Analyst** A person skilled in defining problems for a computer.

**APL** *A P*rogramming *L*anguage. One of the more modern mathematically oriented programming languages.

**Application** The problem for which a computer solution is designed.

**Application program** The software for a computer system may be classified as applications programs and systems programs. An application program is designed to solve a certain type or class of problems. For example, one might have an application program designed to solve a certain type of equation, or to perform a specific statistical computation.

**Array** An ordered arrangement of items.

**Assembler** A computer program that takes instructions written in assembly language and converts the instructions into a language that the computer understands, machine language.

**Assembly language** A computer language intermediate between machine language and compiler languages. It allows machine language instructions to be written in simplified form using mnemonics and other standardized abbreviations.

**Auxiliary operation** An operation performed on equipment not under the direct control of the central processing unit.

**Auxiliary storage** A storage that supplements the internal storage of a computer.

**BASIC** *B*eginner's *A*ll-purpose *S*ymbolic *I*nstruction *C*ode. A programming language that is ideally suited for use in secondary schools. It is one of the easiest languages to learn, to use, and to teach. It has been implemented on many minicomputer, microcomputer, and time-sharing computer systems.

**Batch processing** Handling programs by grouping them into batches.

**Binary** A numbering system based on two values or a system to the base 2.

**Binary digit** A digit in the binary scale of notation. This digit may be a 0 (zero) or a 1 (one). Abbreviated as *bit*.

**Bit** A binary digit.

**Bug** An error in a computer program.

**Branch** A technique used to transfer control from one sequence of a program to another.

**Byte** A group of adjacent bits operated upon as a unit. Most often consists of eight bits.

**Card** A storage medium in which data are represented by means of holes punched in vertical columns in a paper card.

**Card reader** An input device that transfers the information on a punched or mark sense card into the computer.

**Cassette** A magnetic tape storage device. A cassette consists of a tape housed in a plastic container.

**Cathode ray tube** (CRT) An electronic tube with a screen upon which information may be displayed. A keyboard terminal containing a cathode ray tube is often called a CRT terminal.

**Central processing unit** The components of a computer system that contains the arithmetic-logic unit and control circuits. Commonly called the computer.

**Character** A letter, digit, punctuation mark, or other symbol used to represent information. Computers are designed for the input, storage, manipulation, and output of characters.

**Chip** A small integrated-circuit package containing many logic elements.

**COBOL** *CO*mmon *B*usiness *O*riented *L*anguage. A computer language designed mainly for programming business applications. This is one of the most widely used compiler languages.

**Code** A set of symbols and rules for representing information.

**Coding** Writing computer instructions in a programming language for acceptance by a computer.

**Compiler** A computer program (i.e., software) that translates a program written in a symbolic language, such as BASIC, COBOL, or FORTRAN, into the language of the computer (machine language).

**Compiler language** A language such as BASIC, COBOL, APL, or FORTRAN, designed to assist a computer user in writing procedures to solve problems.

**Computer** See Digital computer.

**Computer art** Art form produced by computers.

**Computer program** See Program.

**Computer science** The field of knowledge embracing all aspects of the design and use of computers.

**Computer system** A digital computer, its related peripheral equipment, and associated software.

**Computing** The act of using computing equipment for processing data.

**Conditional transfer** An instruction that may cause a departure from the sequence of instructions being followed depending upon the result of an operation or upon the settings of an indicator. A variation of control based on some condition.

**Console** The part of a computer system that enables human operators to communicate with the computer.

**Control unit** A unit or portion of the hardware of a computer that is designed to direct a sequence of operations, interpret coded instructions, and send proper commands to the computer circuits.

**CPU** *C*entral *P*rocessing *U*nit.

**Data** A representation of facts or concepts in a formalized manner suitable for communication, interpretation, or processing by people or by automatic means. Information to be processed by a computer program.

**Data bank** A stored collection of the libraries of data that are needed by an organization to meet its information processing and retrieval requirements.

**Data preparation** The process of organizing information and storing it in a form that can be input to the computer.

**Data processing** The generic term for operations performed with automatic equipment, which may or may not include a computer. Synonymous with information processing.

**Debugging** The process of eliminating mistakes from a flowchart or computer program, or malfunctions from a hardware device.

**Decision** The computer operation of determining if a certain relationship exists between words in storage or registers and of taking alternative courses of action as a result.

**Digit** One of the symbols of a number system used to designate a quantity.

**Digital computer** A machine that processes information represented by combinations of discrete data. It is capable of accepting information, applying prescribed processes to the information, and supplying the results of these processes.

**Direct access** Pertaining to storage devices where the time required to retrieve data is independent of the physical location of the data. A computer's main memory is a direct access storage device. A magnetic disk is another example of a direct access storage device.

**Disk storage** A storage device that uses magnetic recording on flat rotating disks. It is a direct access storage device.

**Display** A visual representation of data.

**Documentation** The process of organizing information about a problem into a useful file. Usually includes problem statement, associated algorithms, flowcharts, punched cards, program listings, and program operating instructions.

**Downtime** A time period during which the computer system is malfunctioning.

**EDP** An acronym for *E*lectronic *D*ata *P*rocessing.

**Execution time** That period of time during which a program is running.

**File** An organized collection of related data.

**Firmware** Programmed instructions or data that are stored in a fixed or firm way, usually implemented in a ROM. Software programs are stored on paper or magnetic media and must be loaded into the memory of a computer each time the computer is turned on.

**First generation computer** Computers whose circuitry depended heavily upon vacuum tubes. The vacuum tube era ended about 1958, when transistorized computers began to be produced.

**Floppy disks** An auxiliary storage device used with computers. The floppy disk is housed in a plastic cartridge resembling the cardboard jackets that 45-rpm audio records are stored in.

**Flowchart** A pictorial description of a computer solution to a problem, used as a guideline for the preparation of a computer program. A flowchart consists of symbols such as arrows, rectangular boxes, circles, and other symbols used graphically to represent a procedure or pattern of computation to solve a particular problem.

**Flowcharting symbol** A symbol used to represent operations or flow on a flowchart.

**Flowcharting template** A plastic guide containing cutouts of the flowcharting symbols that are used in drawing a flowchart.

**Flowline** A means of connecting two flowchart symbols on a flowchart.

**FORTRAN** *FOR*mula *TRAN*slator. A computer language designed mainly for programming scientific applications.

**Fourth generation** A modern computer that uses large scale integration (LSI) or very large scale integration (VLSI) circuitry.

**Hardware** The physical equipment of a computer system. Contrast with software.

**Information processing** The processing of data representing information and the determination of the meaning of the processed data.

**Information retrieval** A branch of computer science relating to the techniques for storing and searching large or specific quantities of information.

**Input** To enter information into the procedure or computer.

**Input/Output** A general term for the equipment, data, or media used in the entering or recording function. Commonly abbreviated I/O.

**I/O device** A unit that accepts new data, sends it into the computer for processing (input), receives the results from the computer, and converts them into a usable form (output). Common I/O devices are card readers/punches, printers, display terminals, magnetic disk units, magnetic tape units, and digital plotters.

**Instruction** A command to be executed by the computer. A coded program step in a programming language.

**Integrated circuit** An electronic circuit in which all of the components are fabricated together into one tiny unit.

**Iterative** Describing a procedure or process that repeatedly executes a series of operations until some condition is satisfied.

**Interface** The point of contact between different systems or parts of the same system. A shared boundary, that is, the boundary between two devices.

**Interpreter** A computer program that translates and executes each source language statement before translating and executing the next one. Most microcomputers, minicomputers, and time-sharing computer systems use interpreters to process programs written in the BASIC language.

**K** The letter K is used in computer science to represent the number $2^{10}$, or 1024. The size of a computer's memory is often stated in terms of a number of K words or bytes. Thus a small computer memory might be 4K or 8K words, while a large computer memory might be 96K or 128K words, or more.

**Language** A set of rules and conventions used to convey information.

**Large scale integration (LSI)** An integrated circuit that contains a large number of transistors and other circuitry on a single chip. LSI chips are used in microcomputers, microprocessors, and memories.

**Library** A collection of computer programs that are stored on magnetic disks, magnetic tapes, or paper tapes.

**Line printer** A printer where an entire line of characters is composed and determined within the device prior to printing. A whole line is printed nearly simultaneously; speeds of 200–2000 lines per minute are common, as are line lengths of 80 to 132 characters.

**Loop** A sequence of operations usually repeated a controlled number of times within a procedure.

**Machine language** The language or instruction set that a computer is constructed to understand or be able to perform. The language of the computer.

**Magnetic core** A data storage device based on the use of a highly magnetic, low-loss material, capable of assuming two or more discrete states of magnetization.

**Magnetic disk** A flat circular plate with a magnetic surface on which data can be stored by selective magnetization of portions of the surface. It is considered a direct access storage device.

**Magnetic tape** A device for storing digital data in the form of magnetized areas on a tape of plastic coated with magnetic iron oxide.

**Malfunction** A failure in operation of the hardware of a computer or peripheral unit.

**Memory** The storage area of a computer.

**Minicomputer** A small, low-cost computer. A microcomputer contains at least one microprocessor and can be contained on a board or chip.

**Microcomputer chip** A microcomputer on a chip. Differs from a microprocessor in that it not only contains the central processing unit, but also includes on the same piece of silicon, a RAM, a ROM, and input/output circuitry. Often called a "computer on a chip."

**Microprocessor** The brains of a microcomputer. Usually a large scale integration (LSI) chip containing a central processing unit (arithmetic unit, timing and control unit, general purpose registers, and instruction register). To form a working system, at least one external memory device is usually used with the microprocessor chip.

**Microsecond** A millionth of a second.

**Millisecond** A thousandth of a second.

**Minicomputer** A relatively inexpensive computer that has been designed for use in various application areas.

**Monitor** A program to supervise the proper sequencing of programming tasks by the computer. It is an example of computer software, and is often used synonymously with executive, supervisory routine, and operating system.

**Nanosecond** A billionth of a second.

**Numerical analysis** The branch of mathematics concerned with the study and development of effective procedures for computing answers to problems.

**Off-line** Peripheral units that operate independently of the computer. Devices not under the control of the central processing unit.

**On-line** Peripheral devices operating under the direct control of the central processing unit.

**Output** To transfer information from the computer to some device that converts it to a usable form, such as a hardcopy printout or CRT display.

**Peripheral equipment** Ancillary devices under the control of the central processing unit, such as magnetic tape units, printers, card readers, CRT display terminals, or floppy disk units.

**Picosecond** One-thousandth of a nanosecond; or $10^{-12}$ second.

**PL/1** A user-oriented language introduced by IBM, combining the better features of FORTRAN, COBOL, and ALGOL to provide a universally usable language. It can be used for programming both scientific and business applications.

**Print** To transfer information, usually from the computer's internal storage, to a printing device.

**Problem definition** The formulation of the logic used to define a problem. A description of a task to be performed.

**Processing** A term including any operation or combination of operations on data, where an operation is the execution of a defined action.

**Program** An ordered list of statements that directs the computer to perform certain operations in a specified sequence to solve a problem. To design, write, and test one or more programs.

**Program library** A collection of computer programs.

**Programmer** A person who prepares problem-solving procedures and flowcharts and who writes, debugs, and documents computer programs.

**Programming** The technique for translating the steps in the solution of a problem into a form that the computer understands.

**Programming language** A language used to express programs. BASIC, FORTRAN, COBOL, APL, and PL/1 are examples of common programming languages.

**PROM** An acronym for *P*rogrammable *R*ead *O*nly *M*emory. A memory, used with microcomputers, that can be programmed by electrical pulses. Once programmed, it can only be read.

**Punched card** A card that is punched with combinations of holes representing letters, digits, or special characters.

**Punched tape** A paper or plastic tape in which holes are punched. It serves as a data storage device.

**RAM** An acronym for *R*andom *A*ccess *M*emory. A semiconductor memory used in modern computers.

**Random access** Descriptive of storage devices where the time required to retrieve data is not significantly affected by the physical location of data.

**Read** To sense data from an input medium.

**Real time** Descriptive of on-line computer processing systems that receive and process data quickly enough to produce output to control, direct, or affect the outcome of an ongoing activity or process.

**Record** A set of data pertaining to a particular item. Records are, in turn, grouped together to form files.

**Remote terminal** A device for communicating with computers from sites that are physically separated from the computer, and often distant enough so that communications facilities such as telephone lines are used rather than direct cables.

**Response time** The time it takes the computer system to react to a given input. It is the interval between an event and the system's response to the event.

**ROM** An acronym for *R*ead *O*nly *M*emory. An unerasable, permanently programmed memory usually used to store programming language translators, such as BASIC, or application programs, such as chess. Programs stored in ROM are called firmware.

**Run** The single and continuous execution of a program by a computer with a given set of data.

**Sequence** An arrangement of items according to a specific set of rules.

**Second generation computers** A computer belonging to the second era of technological development of computers when the transistor replaced the vacuum tube. These were prominent from 1959 to 1964, and were displaced by computers using integrated circuitry (IC).

**Simulation** The representation of physical systems and phenomena by computers, models, or other equipment.

**Software** Programs and associated material for use on a computer. Includes computer programs, flowcharts, punched cards, program listings, etc. Contrast with hardware.

**Sort** To arrange data into some predefined order.

**Source language** The language that is an input for statement translation.

**Source program** A computer program written in a source language such as FORTRAN, BASIC, or PL/1.

**Special character** A graphic character that is neither a letter or a digit, e.g., + or ?.

**Standard** An accepted and approved criterion for drawing flowcharts, writing computer programs, etc.

**Statement** In programming, a generalized instruction in a source language.

**Storage** The part of the computer that retains information. Internal storage is part of the central processing unit. Auxiliary storage (magnetic disk, magnetic tape, etc.) devices are used to supplement the internal storage of a computer.

**Storage capacity** The amount of data a storage device is capable of holding.

**Subroutine** A group of statements directing a computer to perform a particular operation that may be used repeatedly in a procedure.

**Symbol** A letter, numeral, or mark that represents a number, operation, or relation. An element of a computer language's character set.

**System** A collection of machines and methods required to accomplish a specific objective. A computer system consists of hardware and software.

**Systems programs** The software for a computer system may be classified as applications programs and systems programs. The systems programs include assemblers, compilers, debugging aids, and operating systems.

**Terminal** A point in a computer system or communication network at which information can either enter or leave. Some terminal devices are CRT display units, digital plotters, and card readers.

**Third generation computers** Computers that use large scale integrated circuitry and miniaturization of components instead of transistors.

**Throughput** The total amount of useful work performed by a computer system during a given time period.

**Time-sharing** A method of operation in which a computer facility is shared concurrently by several users.

**Turnaround time** The amount of time that is required for a computational task to get from the computer user to the computer, onto the machine for execution, and back to the computer user in the form of the desired results.

**Write** The process of transferring information from the computer's memory to an output medium.

# INDEX

ABS function, 61
absolute value, 61
abundant numbers, 128-130
algorithm, 17-19
amicable numbers, 123-126
APL, 22
arithmetic expression, 41
arithmetic operators, 40-41
Armstrong numbers, 126-127
arrays, 70-75
  one-dimensional, 70-73
  two-dimensional, 70, 73-75
ATN function, 60
auxiliary storage, 8-10

BASIC, 17, 24-103
  acronym, 25
  characters, 27
  constant, 28
  expression, 40-41
  functions, 59-68
  line number, 25-26
  names, 29-30
  operators, 40-41
  statements, 25
  subroutine, 59, 68-70
  symbols, 27
  variables, 29-30
binary numbers, 196-197
bubble memory, 8

casting out nines, 208-210
central processing unit, 6
Chinese remainder theorem, 210-213
clock arithmetic, 206-208
comments, program, *see* REM statement
computer,
  applications, 1-3
  central processing unit, 6
    importance of, 1-3
    language, 21-22
    large-scale, 12
    medium-scale, 13-14
    microcomputer, 13-15
    microprocessor, 15-16
    minicomputer, 13-14
    pocket, 25
    small-scale, 13

  storage, 8-10
  supercomputer, 12
  systems, 10-16
computer systems, 10-16
control statements, 46-59
COS function, 60
CPU, 7
CRAY-1 supercomputer, 5, 7, 12, 115

DATA statement, 33-35
decision symbol, flowcharting, 20
decimal numbers, 199-200
DEF FN statement, 67-68
deficient numbers, 128-130
DIM statement, 71, 81
Diophantine problems, 221-222

END statement, 31
ENIAC, 17
Eratosthenes, sieve of, 111-113
even order magic square, 169-175
EXP function, 61
exponent, 28
exponentiational symbol, 40
expression, 41

factoring, 134-148
Fermat's last theorem, 229
Fibonacci, Leonardo, 151
Fibonacci numbers, 151-159
firmware, 11
floppy disk, 9
flowchart, 19-21
flowcharting symbols, 19-20
FOR statement, 52-59
FORTRAN, 22
four color problem, 222-224
functions, predefined, 60-66
  arithmetic, 61-63
  exponential, 61
  trigonometric, 60-61
  utility, 63
functions, user-defined, 67-68

geometric magic square, 185-188
Goldbach's conjecture, 229
GOTO statement, 46-48
GOSUB statement, 68-70
greatest common divisor, 141-144

**247**